Economic Effects of Scale Increases in the Steel Industry

Myles G. Boylan, Jr.

Published in cooperation with
The Research Program in Industrial
Economics of Case Western Reserve University

The Praeger Special Studies program—
utilizing the most modern and efficient book
production techniques and a selective
worldwide distribution network—makes
available to the academic, government, and
business communities significant, timely
research in U.S. and international economic, social, and political development.

Economic Effects of Scale Increases in the Steel Industry
The Case of U.S. Blast Furnaces

9 0 6 6 9

Praeger Publishers   New York   Washington   London

Library of Congress Cataloging in Publication Data

Boylan, Myles.
    Economic effects of scale increases in the steel in-
dustry.

    (Praeger special studies in U.S. economic, social,
and political issues)
    Bibliography: p.
    Includes index.
    1. Steel industry and trade—United States.
2. Blast-furnaces.   3. Economics of scale.   I. Title.
TN704.U5B68   1975   338.4'5   74-14675
ISBN 0-275-09790-0

PRAEGER PUBLISHERS
111 Fourth Avenue, New York, N.Y. 10003, U.S.A.

Published in the United States of America in 1975
by Praeger Publishers, Inc.

Printed in the United States of America

ACKNOWLEDGMENTS

I am indebted to Marvin J. Barloon, Bela Gold, Alexander R. Troiano, William S. Peirce, Gerhard Rosegger, and other members of the faculty at Case Western Reserve University for their valuable comments and criticisms of earlier drafts of this book. Special thanks are due Bela Gold, Timken professor of economics and director of the research program in industrial economics, for his indefatigable willingness to aid me in any way and, specifically, for his timely suggestions and generous financial assistance.

Financial assistance was also provided by the American Iron and Steel Institute through the award of a dissertation fellowship. For this and other forms of assistance, I am indebted to Dr. Bertis E. Capehart, director of the education department of the AISI.

Some of the information used in this study was provided by Armco Steel, Inland Steel, Jones and Laughlin Steel, Republic Steel, United States Steel, and Arthur G. McKee and Company. Dr. David Dilley and Dr. David McBride of the United States Steel Corporation, Mr. John Fischley of Republic Steel Corporation, and Mr. William Collison of Arthur G. McKee and Company were particularly valuable sources of technical and economic information.

Finally, my family, and in particular my wife, Margaret, offered encouragement of the type only they could provide.

A scanning of the industrial horizon during recent years reveals an extraordinary rush toward gigantism in steel mills, electric power plants, ships, petrochemical complexes, and many other sectors of manufacturing, agriculture, transport, and even distribution. This reflects a pervasive belief in the advantages of large operating units that has spread beyond engineers and the management of firms to governments as well—leading many of the latter to join in fostering such tendencies in the hope of revitalizing their lagging industries. Even in the midst of this stampede, however, it would seem prudent for corporate as well as public officials to reexamine the foundations of such beliefs and to undertake more careful assessments of the sources and effects of the apparent scale economies realized in the past and of those anticipated in the future.

Economic theory has long held that increases in scale yield progressively lower average total unit costs—up to some optimum point beyond which they turn upward. But critical analysis reveals this theory to be almost devoid of substantive content, suggesting only that there is some optimal size of production unit for any given technology without specifying what it might be, what determines it, or the magnitude of its presumed cost advantages as compared with larger or smaller units.

And the reputedly hard-headed engineering literature reflects longstanding acceptance of a "rule" that each doubling of capacity tends to require increases in investment of only about six-tenths. Further inquiry reveals, however, that this expectation seems rooted in the simple-minded view that the volume increases more rapidly than the enclosing surface of rectangular, cylindrical, and spherical shapes—and that the capacity of facilities tends to be correlated with their volume, while investment requirements tend to vary with the size of the enclosing surface. Such a relationship may hold, of course, in respect to some kinds of apparatus and facilities, especially in respect to the cost of constructing outer shells such as tanks, furnaces, boilers, pipes, and simple buildings. But fundamental shortcomings narrowly restrict the range of its applicability.

Even construction activities involve growing demands on resources under conditions of increasing size, strain, and deteriorative forces. And large facilities often make disproportionate demands on energy inputs, ancillary equipment, instrumentation, and such. Hence, the

---

Bela Gold is Timken Professor and Director of the Research Program in Industrial Economics at Case Western Reserve University.

rate of increase in construction costs with gains in capacity covers a wide range among processes and industries beyond the supposedly universal value of six-tenths. Moreover, there is every reason to expect such exponential values to change over successively higher ranges of scale. Finally, the engineering literature generally ignores the other components of the total unit production costs on the basis of which scale adjustments presumably are evaluated.

In short, although increases in scale undoubtedly yield substantial advantages in many industrial applications, such expectations may prove unduly optimistic and even misleading in others. What is needed to provide more effective guidance in appraising such alternatives is a deeper understanding of the specific sources and limitations of the changes associated with modifications in scale that engender cost advantages. These usually involve alterations in the nature of processes, in the specialization of equipment and labor tasks, in the qualitative characteristics of inputs, and in the design and mix of products. To identify such sources and to measure their effects, it is necessary to penetrate beneath plant aggregates to analyze every stage in the flow of resources through production processes and to relate changes in the physical aspects of such input-output sequences to attendant cost variations. And to carry out such demanding tasks requires a detailed knowledge of the technologies involved, as well as of the multiplicity of paths whereby these may engender economic repercussions.

This defines the objectives and the very substantial achievements of this study by Professor Myles G. Boylan. He has dug beneath industrywide aggregates for blast furnaces to focus on individual furnaces. And he has dug beneath aggregate inputs, outputs, and costs for individual furnaces to examine the ways in which each component of the pig iron producing process has been altered by major changes in scale. Technological changes are identified and measured before tracing their impact on the qualitative characteristics and costs of resource and product flows. More important, the resulting analytical model has been generalized to permit applications to a wide range of industries in which the capital equipment and the quality of material inputs are the dominant determinants of plant capacity and operating efficiency.

In past studies of economies of scale, academic economists have depended almost entirely on two approaches. The first entails fitting data on inputs to data on output using production functions in which certain crucial aspects of the input-output relationships have been prespecified. The other technique is to estimate a realworld counterpart of the long run average cost curve. Both of these approaches are subject to a number of pitfalls. From a purely theoretical point of view, neither can succeed completely in describing the desired relationships except in excessively restrictive circumstances. From a practical point of view, there are too many extraneous influences whose effects must be removed before these desired relationships can be uncovered.

The approach developed here treats the plant as a cohesive network of interdependent activities. Provision is made for changes over time in the relationships among capital, labor, energy, materials, and output in each activity (as well as in the characteristics of the activities themselves) in response to adjustments in supply and demand conditions and in the scale and techniques of production. Although alterations in techniques or scale often are localized initially within certain activities, they tend to engender adaptive adjustments in wider reaches of the plant's network and hence require the tracing of such effects as well for purposes of thorough evaluations.

The relevant input-output relationships are developed in a dynamic sense for each of the basic inputs (materials, energy, capital, and labor) by making extensive use of findings reported in the engineering literature and by applying regression analysis. Thereafter, the relative effects of changes in scale, technology, and supply conditions on the cost of production are considered in conjunction with changes in trans-portion costs (and location) and the (derived) demand for iron products. The important roles played by location and demand in determining the speed of diffusion of increases in scale, improvements in technology, and the average quality of materials are developed.

Professor Boylan's model has a number of merits. It incorporates technological improvements directly into the analysis instead of treat-ing them as a shift parameter or a residual. It permits the joint assess-ment of the effects of changes in scale and technology by allowing for the possibility that these are intertwined. Finally, it minimizes the necessity of adjusting for extraneous factors such as differences among plants and (over time) differences in the scope of plant operations and in product mix.

The results of this research have significant implications not only for the steel industry but for the economy at large as it faces the com-bined pressures of accelerating technological progress and of rapidly expanding consumption of natural resources. Major innovations during the last two decades in the preparation of iron ores were found to have had a substantial impact on costs. These innovations, which were the result of economic pressures created by the increasing scarcity of domestic ore supplies of "normal" quality, made it economically attrac-tive to mine from ores that were far below the average quality that pre-vailed during the first half of this century. Since huge reserves of these low quality ores exist in the United States, the domestic industry has achieved potential self-sufficiency far into the future. Also, the in-creased costs of ore preparation were less than the costs saved in the blast furnace activity as a result of improvements made in key character-istics of the iron materials consumed by blast furnaces. These improve-ments permitted major savings of capital, labor, and energy. Techno-logical improvements in equipment design that accompanied the improved quality of iron materials increased enormously the potential scale of blast furnace operations.

The solution achieved by the steel industry to the problem of critical shortages of iron ores and impending shortages of coking coal of adequate quality merits attention by the large number of corporate and government officials and analysts wrestling with the future implications of current shortages of basic minerals and fuels. The key finding that major innovations can develop in response to economic pressures may provide some reassurance to those concerned with the efficacy of development and innovational programs. Finally, the potential savings in cost attributable to large scale operations in the blast furnace sector should interest antitrust analysts and agencies concerned with the possible need to adapt prevailing policies and restrictions on mergers and joint operations to new technological and scale potentials.

With respect to the specialized interests of economists, Professor Boylan's study contributes to filling the serious gap between theoretical and empirical research on production. It formulates the compromises that have to be considered when attempts are made to apply broad theoretical models to actual operations involving difficult measurement problems and complex interactions. Particularly, it provides a methodology for probing the interactions of technological potentials and economic criteria at the level of intraplant detail where actual alternatives are evaluated, where decisions to adopt or reject are made, and where pressures for adaptive readjustments in adjacent sectors of operations are engendered.

This study complements other research* into the sources, nature, and effects of improvements in technology and productivity in a wide array of industries by the Research Program in Industrial Economics of Case Western Reserve University. Other central research focuses of this program include (1) adjustments in industrial costs, prices and investment; (2) factors affecting growth, fluctuations, and geographical shifts in firms and industries; (3) changing patterns of domestic and foreign competition and integration, including altered relationships between public and private sectors; and (4) the design of more effective managerial forecasting, planning, and control systems.

---

*Other recent volumes include B. Gold, Explorations in Managerial Economics: Productivity, Costs, Technology and Growth (New York: Basic Books, 1971); S. Eilon, Management Control (London: Macmillan, 1971); D. A. Huettner, Plant Size, Technological Change and Investment Requirements (New York: Praeger, 1974); and Technological Change: Economics, Management and Environment, B. Gold, ed. (New York: Pergamon Press, 1974).

CONTENTS

LIST OF TABLES

xvii

LIST OF FIGURES

1

CONCEPTS AND
SOURCES OF
ECONOMIES OF SCALE

This chapter has two basic objectives: to examine critically the prevailing concepts of economies of scale and returns to scale and to review prevailing theories concerning the sources of economies of scale.

A CRITIQUE OF PREVAILING CONCEPTS

Increasing returns to scale are defined to occur over a specified range of homogeneous output when a proportionate increase in all inputs results in a proportionately greater increase in output. Similarly, economies of scale are defined to prevail over a specified range of homogeneous output when a small increase in output in that range results in a proportionately smaller increase in total cost, that is, when long run average cost declines as output is expanded.

Long run cost and economies of scale may be defined to include only the costs of production at the level of the plant—the primary focus in this study. The long run is defined to be a period long enough that firms can adjust all resources to the desired levels and proportions.*
Thus, long run cost would be approximated in a newly built plant designed to meet expected market conditions.

---

*This requirement usually would be dictated by the period of time needed to make and implement new decisions about capital facilities. The long run is not associated with a definite passage of time because it was devised to facilitate static equilibrium analysis; that is, the adjustment process is of only secondary interest; and the focus is on the new levels and proportions of inputs consumed and produced by plants in the industry.

The scale of production generally is accepted as synonymous with plant capacity. But this raises the question, will expanding plant capacity by, say 100 percent, represent the same increase in scale whether it involves duplicating the original production unit or designing a new unit from the outset to have double the capacity of the original unit. Additionally, since modern plants rarely produce a single homogeneous product, plant expansion may involve the addition of various facilities that merely increase the variety of products being manufactured. Because such alternatives are likely to have dissimilar effects on the relationships between inputs and output as well as on the level of average total cost, Bela Gold has offered the following revised concept of scale,

> Specifically, the scale of production may be defined as relating to the level of planned productive capacity which has determined the extent to which specialization has been applied in the subdivision of the component tasks of a unified operation. [1]

Thus, increases in plant capacity need not insure an increase in the scale of production. Indeed, scale could be increased merely by altering the organization, allocation, and utilization of plant resources to concentrate more heavily on a reduced variety of products. This would follow, because in his view,

> [The] limits of specialization tend to be determined in large measure by the level of productive capacity, inasmuch as each separate task must be performed with sufficient frequency to fully occupy the men and equipment assigned to it—otherwise these resources would have to be assigned to several tasks in order to escape the losses involved in idleness. [2]

There is a strong similarity between the concepts of returns to scale and economies of scale. Both are based on proportionately greater increases in output than inputs as the scale of production is increased. The distinction between them is that returns to scale are based on a physical measure of the inputs, necessitating that the inputs be maintained in constant proportions to determine if the relative gain in output is greater than the relative increase in inputs. Since economies of scale are based on a summary financial measure of the inputs in which the inputs' prices are used as weights, the input proportions may be varied without undermining the applicability of this concept. Thus, economies of scale are the less restrictive concept.

The major weakness in the definition of returns to scale is its dependence on fixed input proportions. This means that, in general, the range of output over which returns to scale are found to prevail will depend on the specific proportions selected. Indeed, the fundamental issue is that it may not be possible to increase the scale of production by moving to successively larger plant sizes without altering the input proportions; and even should it prove to be possible to do so, it is doubtful that the fixed proportions would be in accord with managerial preference. Only in the highly limited circumstance in which success- ive increments to plant capacity were achieved by duplicating pro- duction units (which is not an increase in the scale of production) would one expect fixed input proportions to be both possible and desirable.

It may be surmised that the concept of returns to scale has been found useful in theoretical economics primarily because of its strong association with another concept—the production function. In general, the production function is a mathematical expression that equates various amounts and combinations of inputs with various levels of homogeneous products. It represents a summary of the choices con- fronting a firm that wishes to enter the product market by constructing or acquiring a physical plant.* If no restrictions are placed on the production function other than to assume it is continuous and differ- entiable, then the choice of input proportions is independent of the scale of production, because no overt link is present between scale and these proportions. [3] Hence, under a condition of constant input prices, it cannot be claimed that managers would choose different input proportions at larger scales of operation[†] and the requirement that these proportions be fixed to test for returns to scale does not seem excessively restrictive.

In recent decades, two particular types of production functions that reinforce the separation of scale and input proportions have been developed and employed in a large number of empirical studies. These include the Cobb-Douglas function, named after its originators and developed in 1927, and the constant elasticity of substitution function (commonly called the CES function), [4] developed in 1961. These functions lend themselves to empirical application because certain crucial properties have been prespecified. In particular, the flexi- bility that managers have in substituting capital for labor (or the other

---

*More formally, let $V = (v_1, \ldots, v_N)$ represent the vector (array) of distinct inputs and Q the homogeneous product. If a mapping (functional relationship) F exists such that $Q = F(V)$, then F is defined to be the production function.

†To be fair, those employing the general production function in theoretical research do not claim that the same input proportions would be selected, either. It is not possible to tell what input proportions will be optimal at any scale of operations until the production function is made more deterministic; that is, until some of its properties are prespecified.

way around) is assumed to be independent of scale, * and the same degree of returns to scale is assumed to prevail over the entire output range, regardless of input proportions.†Although these functions may be useful as textbook illustrations and as guidelines in empirical macroeconomic studies, their usefulness in empirical microeconomic studies is hampered severely by the limited applicability of these prespecified characteristics.

Economies of scale suffer from a weakness endemic to comparative static concepts. By ignoring the behavior of the <u>components</u> of the total cost of production over time, this concept implicitly assumes that either they will not change or that changes will be mutually offsetting, leaving the long run average cost for any rate of production invariant to time.‡ If the physical plant is an indivisible complex, § it is likely that the variable costs of production will rise over time as the plant ages, because maintenance and repair expenses generally rise. Thus, if the long run average cost is to be invariant to time, it is necessary that the interest and depreciation charges decline over time to offset the rise in maintenance and repair expenses.

In a long run static equilibrium (defined in this study to be a state in which technological progress in the economy is dormant and consumers' preferences are unchanging), the prices of products will be stabilized at fixed real values. In a perfectly competitive industry, these prices will be equal to their respective long run average costs. Hence, interest and depreciation charges must fall over time as maintenance and repair costs rise, as discussed in the previous paragraph. A detailed examination of these fixed charges indicates that under no circumstances will the true time profile of depreciation

---

*In the Cobb-Douglas case, the technical expression for this flexibility—called the elasticity of substitution—is assigned the value one. In the CES case, the elasticity of substitution may be assigned any positive value <u>except</u> one. A value of one means that if the price of labor increases by 50% relative to the price of the services of capital, then the capital input will increase 50% relative to the labor input. A higher value means that more extensive substitution of capital for labor would be undertaken, given the same 50% change in relative prices.

†Production functions that possess this property are said to be homogeneous. Homogeneity of degree one means that constant returns to scale prevail. If a function is homogeneous of a degree greater than one, then increasing returns to scale are present.

‡ This statement abstracts from changes in the general price level in general, and changes in input prices in particular. While these changes will affect long run average cost in nominal terms they will not affect it in real terms.

§ It is not essential to the following argument to make this assumption. It merely facilitates the exposition to do so.

charges* correspond to those determined by conventional accounting methods (such as straight line, sum of the digits, and declining balance formulas).[5] This leads to the conclusion that even while perfectly competitive, long run equilibriums are established, it will not be possible to determine the correct long run average cost of production from accounting data, and thus the extent to which economies of scale prevail cannot be measured accurately.

In static equilibrium in an imperfectly competitive market or dynamic equilibrium in a competitive market, the long run average cost will change over time, either necessitating a modification of the static long run average cost conception or a careful adjustment for the variance in plant vintages and technology. Additionally, the true depreciation charges do not correspond to their accounting counterpart under a variety of reasonable assumptions about the relationship and behavior of a firm's product prices in these kinds of equilibriums.†

This concern with the behavior of long run costs, particularly capital costs (depreciation and interest charges), is not mere "nit-picking." First, either the concept of economies of scale must be recognized as being dependent on the assumption that the plant supplies products to a competitive market in static equilibrium-conditions unlikely to be fulfilled in most manufacturing industries—or it must be made explicit that long run average cost is not invariant to time. Second, the finding that annual capital costs are not reflected correctly under prevailing accounting practices—even if the concept of economies of scale is totally applicable—implies that statistical efforts to measure accurately long run average cost will be defeated in the best of circumstances. Since most empirical studies uncovering economies of scale find only modest savings in the high output range (for example, a two per cent reduction in average cost over the top quartile of plants), the deviation of the true from the reported capital

---

*The true depreciation in the value of a fixed asset is measured by the decline in the present value of future streams of net income (total revenue-variable cost).

†To prove this statement, it was convenient to assume that in equilibrium in imperfectly competitive markets the net rate of entry of firms into the market was zero and each firm's prices remained in constant (or predictable) relationship to the prices charged by its rivals. For competitive equilibrium to be compatible with changing product prices and long run average cost, it is necessary that managers be able to predict price changes (or else come to expect them through a mechanism of adoptive expectations).

costs does not have to be large before such findings become value-less.*

SOURCES OF ECONOMIES OF SCALE

Indivisibilities and Specialization

Past attempts to explain the superior efficiency of large scale production techniques seem to fall into two camps. Some (for example, Frank Knight, Nicholas Kaldor, Abba Lerner, and to some extent Joan Robinson and William Baumol)[6] cite indivisible inputs as the primary source of such economies. Others (for example, Adam Smith, Edward Chamberlin, Bela Gold, and to some extent Joan Robinson)[7] claim there are increasingly more specialized techniques of production that can be employed only as the scale of production is increased.

Historically, there has been some confusion attached to the precise meaning of indivisible inputs and, hence, some uncertainty about the relationship of indivisibilities to increased specialization. Baumol, for example, notes that "it is tempting to argue" that production functions exhibit constant returns to scale, because (given homogeneous inputs and constant input prices) it is the view of "some economists" that the only logical reasons for nonlinear production functions are (1) the possibility that at least one input is limited in supply, and (2) indivisibilities.

> Some inputs just do not come in small units. We cannot install . . . half a locomotive (a small locomotive is not the same as a fraction of a large locomotive). As a result, only if operations are carried on a sufficiently large scale, will it pay to employ such indivisible items. This, it is said, is the only source of economies of large-scale production. In other words, from this point of view all production functions are linear and homogeneous, only, unfortunately, it is not always possible to increase or diminish all inputs in exactly the same proportion. [8]

Baumol refuses to succumb to this temptation despite his reasonably clear explanation of indivisibilities and concludes that linear production functions are a matter for empirical investigation, not unsupported conjectures.

---

*Only if the choice of scale does not involve a choice of widely dissimilar inputs that are variable in the short run (such as materials and labor) and only if the average variable cost (in the short run, given current input prices) and average fixed cost (based on a particular depreciation formula assuming a plant life of X years) are both less at a larger scale of operations, does one have reasonable assurances that the large scale plant will be more efficient in the future.

The confusion attached to the indivisibilities concept arises from the standard assumptions of production theory that inputs are both perfectly divisible and homogeneous. With respect to the capital input, these twin assumptions cannot be fulfilled jointly whenever there is the slightest possibility of either varying the sizes of individual pieces of equipment comprising the physical plant or switching to a different, larger scale technique of production. [9]

The assumption that inputs are homogeneous and divisible is not an obstacle for the abstract production function of economic theory, because an abstract unit of measure for capital (and other inputs) can be postulated and the production function still can fulfill certain theoretical roles. Nor does it present difficulties in empirical analyses of production functions at the macroeconomic level, where the capital input is approximated by the investment expenditures associated with it—thus sidestepping the need for more precise capital measurement.

But serious problems are confronted in efforts to fit production functions to empirical data, including direct measures of the capital stock, at the plant and even industry level. These measurements problems should not be interpreted as being indicative that production processes that can be implemented on a larger scale are atypical of production processes in general; the evidence points in the opposite direction.

Joan Robinson has given a clear view of these measurement problems and the general characteristics of indivisibilities, including their nature as one species of increased specialization.

> If all the factors of production were finely divisible, like sand, it would be possible to produce the smallest output of any commodity with all the advantages of large scale industry. But actually, the factors consist of men, money capital, which is finely divisible, like sand, but must be turned into instruments of production each of which, for technical reasons, must be of a certain size; and land, which is usually divisible, but which sometimes, for technical reasons, cannot be divided without limit.
>
> * * * *
>
> Men differ in their natural abilities, and can acquire skill when they concentrate on one single task;[1] (1. The increase in efficiency arises from the fact that 'practice makes perfect'); acres vary in their natural capacities, and machines can be designed for special tasks. For any kind of production there will be a hierarchy of possible technical methods each using more highly specialized units of the factors than the last, and production is carried out most efficiently when each separate action in the productive process is performed by a unit of a factor of production specially adapted (by nature, by practice, or by human ingenuity) to that particular task. But since the (specialized)

units of the factors are indivisible, the most specialized
method of production involves the largest outlay, and it
is not profitable to make use of the full equipment of highly specialized factors for a very small output. As output
increases a method higher in the hierarchy of specialization
can be adopted, and for this reason cost falls as the output
of a commodity increases.

* * * *

. . . in every case where increasing returns (to scale) are
found there must be some point in the process of production
at which a single unit of a factor is engaged. [10]

In regard to the last passage, if Joan Robinson were to analyze operations in an integrated steel mill that had six coke ovens, two blast
furnaces, four open hearth steel furnaces, and one rolling mill, the fact
that only one rolling mill was present would represent deductive evidence
that the mill was in the region of economies of scale. On the other hand,
if the same steel mill were to operate two rolling mills (of the same
type), then Joan Robinson would conclude that the mill was larger than
the minimum efficient plant size.

The primary difficulty with the indivisibilities concept is its
superficiality—its avoidance of detailed analytical descriptions. It is
this seeming simplicity that has led some economists astray, causing
them to conclude that if the indivisible factors were divisible, there
would be no fundamental basis for economies and diseconomies of
scale, reducing production theory to a theory of optimal input proportions. This semantic confusion was challenged by Chamberlin, who
stated,

A meaningful and realistic way to achieve continuous
divisibility of a factor is to change it qualitatively.

* * * *

While human beings are diverse by nature and training,
capital instruments are so by manufacture.

* * * *

It would appear that continuous divisibility in the capital
factor would be completely general were it not for economies in the production of capital instruments themselves
through concentration of a limited number of models. [11]

Chamberlin stresses that the optimal proportions can vary with size.
A correct statement of the indivisibilities concept—such as Robinson's—also recognizes this is true.

Chamberlin also emphasizes that it is increased specialization
that represents the fundamental source of economies of scale. His
view is that the qualitative changes in the factors of production as
the scale of operations is increased may be large, [12] and thus

analogies to big locomotives versus small locomotives are too
simple in nature to constitute a completely representative example
of the evolution to large scale techniques of production. For example,
a doubling of the capacity of an assembly line could involve a change
from a medium speed line jointly dominated by capital and labor[13]
to a high speed highly automated line that bears little resemblence
to its predecessor. On the high speed line the kinds of equipment are
more complex and specialized, and the tasks assigned to many of
the workers are narrower and more repetitive in terms of the range of
skills and effort required.

In summary, there is little question that the possibility of increasing the size of a locomotive, blast furnace, or similar equipment
in which volume plays a role in determining its capacity constitutes a
source of economies of scale that is attributable to the presence of
an indivisible input. But whether the (above) assembly line example
represents this species of specialization is debatable and is more a
question of semantics than content. An economist in the style of Chamberlin would claim that the inputs constituting the medium speed line
(particularly capital) were qualitatively divisible (and thus that the
evolution to the high speed line represents increased specialization).
But an economist in the style of Robinson could claim that, although
the first stage of capital—money capital—used to build the medium
speed line was divisible (because it is homogeneous), the line itself
was indivisible, thus indicating a potential to achieve economies in
a larger scale line.

The Power Rule

The power rule, often used by engineers to estimate costs for
prospective new plants, supplies one possible explanation why one
large capital unit could be preferred to two smaller units. Although
the origins of this "rule" are in doubt, it can at least be traced back
to a 1947 article by R. Womer (and it probably was formulated much
earlier), in which he suggested that capacity is proportional to volume,
and construction cost is proportional to the enclosing surface area.[14]
Perhaps the best statement of this explanation has been made by
F. Moore.

The (power) rule has been adduced from the fact that for
such items of equipment as tanks, gas holders, columns,
compressors, etc., the cost is determined by the amount
of materials used in enclosing a given volume, i. e., cost
is a function of surface area; while capacity is directly
related to the volume of the container. Consider a spherical
container. The . . . cost varies as the capacity (raised)

to the 2/3 power . . . . If the container is cylindrical, then . . . cost varies as capacity (raised) to the 1/2 power, if the volume is increased by changes in diameter, and if the ratio of height to diameter is kept constant, cost varies as capacity to the 2/3 power. [15]

Moore's evaluation of the power rule is sympathetic.

Originally the .6 rule was applied to individual pieces of equipment or processes. A reasonable case can be made for its validity in those cases; however, the . . . formula . . . cannot be indefinitely extrapolated. There are several reasons for this. In the first place, an extrapolation . . . may lead to sizes of equipment which are larger than the standard sizes available or into stresses beyond the limits of the material involved . . . . Second, in some industries expansion takes place by a duplication of existing units rather than by an increase in their size. If the rule is to be applied at all it is safest to limit its use to the range of capacities found in observations.

The .6 rule when applied to complete plants runs into difficulties not encountered on individual equipment. Some expenditures are relatively fixed for large ranges of capacity, . . . complicated industrial machinery does not necessarily exhibit the same relationships between area (cost) and volume (capacity) as do simple structures like tanks and columns. [16]

S. Shuman and S. Alpert, both chemical engineers, are more cautious than Moore in admitting the significance of the power rule. In commenting on the Moore article, they state,

Various reasons may be brought up as to why large plants cost less per unit of potential productive capacity than smaller ones (such as the surface area—volume relationship . . .). But these are only limited explanations, rather than proofs. There are obvious difficulties in defining or circumscribing both 'plant' and 'cost.' Even if the terms could be circumscribed to some extent, the cost of a plant is the summation of many components so diverse that to expect them to follow the same mechanics would be absurd. In the face of these difficulties, one must conclude that any relationship between plant capacity and cost would be empirical, limited and highly approximate.

* * * *

Since the exponential term is a priori never known, chemical economists use the power rule rarely, and only as a planning guide. The important point to be made here is that the power rule is of much more importance and significance to general and theoretical economics than to chemical engineering. [17]

The points made by Moore, Shuman, and Alpert can be interpreted on two levels. First, any claim to the universality of the six-tenths rule in manufacturing is clearly wrong. Although the unit capacity cost has been found to decline with size in an impressive variety of production processes, the power value is not always six-tenths. It can be as low as four-tenths or as high as one. Furthermore, the power value can change over the range of process capacities actually installed. Estimation of a particular power value through multiple regression analysis effectively estimates only the average value (and is biased toward the power value pertaining to average capacities). Therefore, although the power rule may be employed reasonably for some types of individual equipment units, extending its application to the plant level has no discernible basis in economics or engineering.

Second, although no specific power value can be employed as an accurate predictor of construction costs, widespread empirical findings of values significantly below one represent hard evidence that increases in scale often may be expected to yield savings in these costs. As an example of such findings, C. Dryden and R. Furlow have estimated the power value for blast furnaces to be 0.68 for furnace capacities between 1,000 and 3,000 tons per day. [18] In 1967 dollars, their finding translates into a unit capacity cost that ranges from \$32.60 at 1,000 tons per day to \$22.82 at 3,000 tons per day.

Clearly, whenever the power rule can be applied with reasonable accuracy to equipment units or plant processes over a broad range of capacities, there is strong deductive evidence of economies of scale. This follows both because the existence of high capacity units suggests they are more efficient, and because the equipment units and processes susceptible to this kind of construction cost-capacity relationship tend to be of the type where capacity increases exceed increases in the number of hourly workers needed to tend the equipment. As long as the materials and energy inputs do not rise per unit of output, economies of scale are present, since both labor and capital costs tend to decline. (However, Moore's warning that such findings cannot be indefinitely extrapolated still applies.)

In a more formal manner, Alan Walters demonstrated that a power value of less than one for a plant process is equivalent to increasing returns to scale if both the input and output markets are perfectly competitive and a general Cobb-Douglas production function is used. [19] Recalling, however, the severe restrictions imposed on the general nature of production process by this production function indicates Walters' finding must be discounted accordingly. It is also difficult to conceive of perfect competition prevailing in the output market when increasing returns to scale are prevalent in the industry.

SUMMARY

This review has dealt with two fundamental issues. First, although the concept of economies of scale is intuitively clear, the complicated nature of costs and prices invalidates the appealing simplicity of the long run average cost concept. Second, with the limited exception of equipment units where volume plays a major role in determining capacity, attempts to generalize explanations of the sources of economies of scale are vague—and can be confusing as well. This may indicate such sources are specific to the industrial process in question. It certainly calls for further research in this area.

NOTES

1. B. Gold, Foundations of Productivity Analysis (Pittsburgh: University of Pittsburgh Press, 1955), p. 116.

2. Ibid., p. 116.

3. This point was made strongly (although for different reasons) by E. Chamberlin, "Proportionality, Divisibility, and Economies of Scale," Quarterly Journal of Economics 62, no. 1 (February 1948): 229.

4. R. Solow, Minhas, Arrow, and Chenery, "Capital-Labor Substitution and Economic Efficiency," Review of Economics and Statistics 43, no. 2 (May 1961): 225-50.

5. This detailed examination (and the proof of this statement) is contained in M. G. Boylan, "Depreciation, Long Run Cost and Economies of Scale: A Critical Examination," working paper, Department of Economics, Case Western Reserve University 1974.

6. F. Knight, Risk Uncertainty and Profit (New York: Houghton Mifflin, 1921), p. 98; N. Kaldor, "The Equilibrium of the Firm," Economic Journal 44 (March 1934): 65; A. Lerner, Economics of Control (New York: Macmillan, 1974), p. 143; J. Robinson, The Economics of Imperfect Competition (London: Macmillan, 1933), pp. 329-48; and W. Baumol, Economic Theory and Operations Analysis (Englewood Cliffs, N. J.: Prentice-Hall, 1961), pp. 180-81.

7. A. Smith, Wealth of Nations Book I [copy of fifth edition of original] (New York: Modern Library, 1937), ch. 2; E. Chamberlin, "Proportionality, Divisibility and Economies of Scale," Quarterly Journal of Economics 62, no. 2 (February 1948): 229-62; B. Gold, op. cit., pp. 115-17; and J. Robinson, op. cit.

8. W. Baumol, op. cit.

9. According to Edward Chamberlin, op. cit., p. 229, ". . . the erroneous thesis has come to be widely held that under the 'perfect divisibility' of theory, as applies to the factors of production, there would be no economies of scale." Chamberlin traces the inception of this "erroneous thesis" to Frank Knight, and maintains that later

formulations of it appeared in articles by Kaldor and Lerner. For example, Kaldor, op. cit., stated, "It appears methodologically convenient to treat all cases of large scale economies under the heading 'indivisibility.' . . . (It may be) not so much the 'original factors,' but the specialized functions of those factors which are indivisible." Chamberlin's cogent remark about this passage was "Divisibility is thus defined to include the availability for small scales of these 'specialized functions'," op. cit., p. 237. The point to be emphasized is that measurement problems related to the indivisibility of homogeneous capital inputs (or alternatively the heterogeneity of divisible capital inputs) have been extended incorrectly to the general properties of production processes themselves.

10. J. Robinson, op. cit., pp. 334-37.

11. Chamberlin, op. cit., pp. 242-43.

12. He also notes that "the product itself ordinarily undergoes qualitative change, often quite drastic, . . ." Ibid., p. 236, footnote.

13. See: B. Gold, op. cit., pp. 223-24. A jointly dominated process is defined by Gold to be one in which both capital and labor determine the capacity of the process.

14. R. Womer, Chemical Engineering 54, no. 12 (December 1947): 125.

15. F. T. Moore, "Economies of Scale: Some Statistical Evidence," Quarterly Journal of Economics 73, no. 2 (May 1959): 234-35.

16. Ibid., pp. 235-36.

17. S. Shuman and S. Alpert, "Economies of Scale: Some Statistical Evidence: Comment," Quarterly Journal of Economics 74, no. 3 (August 1960): 493-94.

18. C. Dryden and R. Furlow, Chemical Engineering Costs (Columbus: Engineering Experiment Station, 1966), p. 71. This source contains estimates of power values for thousands of equipment units and many plant processes. The majority fall in the range from 0.45 to 0.90.

19. A. Walters, "Economies of Scale: Some Statistical Evidence: Comment," Quarterly Journal of Economics 74, no. 1 (February 1960): 154-57.

One of the key tests of an economic concept is the ease with which it can be applied to further our understanding of economic events. For example, if an industry is dominated by a few large firms, the concept of economies of scale can be offered as an initial hypothesis to explain the dominance by large firms. Unless this hypothesis can be accepted on statistical grounds, however, the value of this concept is limited to classroom gymnastics. Statistical analysis requires that inputs or cost be related properly to output at different scales of operation while controlling for the effect of extraneous factors. A number of such techniques have been devised to explore the possible existence of economies of scale. In the following three sections, these methods are critically reviewed.

DIRECT EVIDENCE ON ECONOMIES OF SCALE: THE STATISTICAL AND ENGINEERING APPROACH

What direct evidence is there on economies of scale? In a 1955 survey article, Caleb Smith attempted to assess the extent to which economies of scale prevailed in business.[1] The tone of his article was pessimistic about the likelihood that economists ever will be able to make this assessment effectively. He suggested, and his colleague Milton Freidman concurred, that perhaps economists are asking the wrong question when they focus their interest on the long run average cost curve. This suggestion has been made by others, notably by Bela Gold.[2] Some of Smith's remarks are illuminating:

The effort to obtain empirical evidence of relationships which have been developed in deductive economic theory encounters two sets of problems:

14

1. The nature of empirical facts—we are considering empirical facts which have meaning only in the context of a complicated theoretical framework.

2. Knowing when to eliminate variations in cost which are not a result of economies of scale.

Simplifying assumptions are essential to the development of theoretical concepts. Inevitably, however, each simplifying assumption blocks the path toward empirical investigation of the relationship which the theory states. [3]

Smith considers the measurement of "size of plant" (or firm) to be "unequivocal only if output is homogeneous." [4] He cites the well-known accounting problem of allocating plant (or firm) costs among multiple products. When a firm expands from a small size to a large size, the output rarely remains of a single type. It is rare to find widely different plant sizes producing a homogeneous product. Even if small and large firms can be found that produce the same product mix, the likelihood is great that cost accounting techniques employed by the firms are significantly different (cost accounting techniques are not as standardized as income accounting techniques).

Smith recognizes two general techniques that can be employed to handle cost variations from causes not directly related to economies of scale, thus resulting in an estimated cost-scale relationship equivalent to the long run average cost curve. In the first approach which he calls the statistical approach,

. . . the costs of the plant or firm as a unit or the costs allocated to some type of output are related to size. Other influences on cost are either ignored or allowed for such techniques as deflation or multiple correlation. [5]

The major weaknesses of this approach, other than the problems of defining size noted above, stem from the fact that costs may be affected by technological progress that is not related to size. For example, the age distribution of the capital stock may not be related to size. Smaller plants may be rebuilt to take advantage of new developments in the array of production techniques, while larger plants that have enjoyed the advantage of economies of scale for a number of years may not feel the cost pressure to rebuild. Inefficient plants may be sold (in some cases, after they have gone into receivership) for a price that reflects a sizeable discount from book value and little premium for good will. If the new owners operate the plant, the fixed cost component will be diminished, making it appear that the plant is more efficient than it actually is.

There are also major difficulties in the way one successfully applies statistical techniques to uncover the relationship between plant size and average cost. Variations in cost due to regional differences in input prices are difficult to remove without exploring the production processes themselves. The largest number of observations tend to be concentrated in the range of plants of intermediate size, making it difficult to identify average cost at small and large scales of operation, with the result that the findings of a particular study often depend on the particular functional form of the estimated regression equation. Thus, if the statistical approach is discounted because of the potential biases in the data and statistical techniques, then little weight can be given to the findings of such cost studies. An excellent summary of the results and estimation problems associated with this approach can be found in Johnston. [6]

In the second approach, which Smith calls the engineering approach,

> Each element of the production process is studied to discover the relation between inputs and outputs at different scales for that process. The input-output relations of the processes are then combined to give the overall input-output relations. The introduction of prices for the inputs transforms these relations into cost-output relations. [7]

This approach is also cited by Gold as one of the more common approaches to be found in estimates of cost savings attributable to larger scale operations or technological change. [8] Although Gold is willing to admit to the reasonableness of the approach in achieving accurate estimates of costs before the new scale or technology enters the diffusion process, he notes that, ultimately, the new process will have effects in both the supplying markets and the output markets that will change the prices of inputs and possibly lead to alterations in the production process. Thus, to the extent that large or small plants use different inputs, or the same inputs in different proportions, estimates made at one point of time in the diffusion process are highly temporal and are subject to change. Smith cites similar weaknesses in the engineering approach, including the narrowness of the input-output investigations, and the ideal rather than actual estimated relationship between cost and size. Overall, Smith believes that the relationship between cost and size is more effectively estimated at the plant level by the engineering approach, because—compared with the statistical approach—this procedure has the advantage of employing the most up-to-date technology at all plant sizes, giving some assurance that estimated average cost differences are not due to systematic differences (biases) in the technologies employed in small and large plants.

A representative example of an engineering cost study is J. Bain's classic study of 20 manufacturing industries, in which he used data from questionnaires to make cost estimates of the minimum efficient plant size. [9] Bain found the long run average cost of production at the plant level to be either flat or L-shaped* in 11 industries in which these determinations could be made. He did not explain (or was not able to explain) why long run average cost did not decline beyond critical plant sizes in these industries, but one may suspect certain limitations were present in his data-gathering technique (questionnaires) and in the data itself (for example, it is possible that in some cases the larger plants merely duplicated facilities found in smaller plants).

Beyond these usual criticisms of data and technique, the (perhaps) more critical issue may be raised concerning the limited ability of engineers (or anyone else) to assess the effects of major increases in scale beyond those already in operation. To successfully predict the minimum efficient plant size, given the level of technological knowledge and the relative prices of inputs, it would seem necessary that scale be susceptible to analytical modeling (for example, fundamental engineering relationships). An example of the infrequent attempts to express the fundamental relationships in the production process in a mathematical model can be found in a study of pipelines by A. Chenery.[10] His approach seemed promising, but is excessively restricted to those processes where; (1) output is homogeneous; (2) capital can be specified in a few dimensions, for example, weight, length, or volume; and (3) the labor requirement is a direct function of the capital facilities. According to Chenery,

> Industry studies have generally used statistically determined cost curves. Since these curves are based of necessity upon productive combinations which it has proved feasible for entrepreneurs to try out, they cannot usually tell us much about the broader range of productive possibilities which have been explored experimentally but not adopted commercially. The lack of this information prevents quantitative analysis of the possibilities of substituting one factor for another, since statistical data only pertain to historical observations in which the effects of technological change and price variations are usually inseparable.[11]

The lack of information on alternative combinations of inputs noted by Chenery, however, generally is due to an incomplete understanding of the fundamental nature of the transformations effected in the plant.[12] For the engineers to estimate efficient combinations, it is often necessary for them to make a series of least cost calculations; otherwise their work would mushroom. One may question their ability to

---

*So-called, because the average cost curve was found to decrease and then level off for increases in scale.

arrive at the most efficient plant design since the danger of suboptimizing is great. But more important, it seems unlikely that engineers can judge the practical technical options that would be efficient under different price conditions from those observed.

INDIRECT EVIDENCE DERIVED FROM DEDUCTIVE MODELS

The Survivor Technique

Following the publication of the survey article by Caleb Smith, George Stigler authored a 1958 article in which he applied a new kind of analysis—he called it the survivor technique—to the problem of determining the extent of economies of scale at the level of the firm in the steel, automobile, and oil-refining industries. [13] Others (for example, T. R. Saving) have used this technique to make similar determinations at the plant level. [14]

This technique is a clever attempt to avoid the pitfalls of using cost data to ascertain the effect of economies of scale on industry structure. It is based on the assumption that the more efficient firms or plants [15] will have a higher probability of survival than the others (particularly when competition is stiff); and as a consequence, the size distribution of firms or plants will tend to move over time toward, or center on, the range of production where long run average cost is at a minimum. Unfortunately, it is really a case of substituting new problems for old.

The first major difficulty encountered by this technique is its assumption that the typical industry market is competitive enough to insure that a strong cost discipline will be exerted on the actions of the constituent firms throughout the country. Related problems are lack of product homogeneity within the industry boundaries and different degrees of product differentiation by member firms (achieved by advertising and other promotional expenditures), a dissimilar age distribution of capital stock across the industry's plants, significant differences in regional prices and markets, rapid changes in technology that may have changed the size of an efficient plant faster than the typical firm's ability to react to the new technology, changes in relative input prices, and an uneven (for example, bimodal) size distribution of firms or plants within the industry due to systematic differences in product specialization and firm size.

As an example of the application of this technique, in a 1961 study by T. R. Saving, [16] based on data from the 1947 and 1954 editions of the Census of Manufactures, 68 of 200 industries randomly chosen for analysis had to be discarded because of obvious lack of product homogeneity, an ambiguous definition of the industry, or because the

Census of Manufactures had to lump a significant number of plants together to avoid disclosure of individual companies. In the latter case, it may be surmised that many highly concentrated industries were dropped from the analysis. Forty-three additional industries were discarded because,

> For the estimation technique to yield unambiguous results it is necessary that the increasing size classes (of plants) lie in a continuous group. However, we occasionally find industries which have two or more distinct groups of size classes with increasing relative shares.[17]

In this case, it is not known whether the cause is random fluctuations, an increase in the range of optimum size, the existence of two ranges of optimum size, improper industry definition, or failure to control for sources of variation noted above.

Aside from these visible deficiencies in the data, it may be surmised that many disrupting influences of the type noted above underlie the changes in the distribution of plants in the remaining 89 industries, greatly detracting from any conclusions Saving was able to draw concerning the concentration of plants in them.

In general, little or no insight is gained from studies employing this technique. Since the inputs and costs are ignored, the survivor technique cannot provide quantitative measures of scale economies; and, given the inapplicability of simple microeconomic models to analyses of most industries, it is too much to expect that changes in industry structure can be explained through such models.

The Prespecified Production Function Approach

A technique that is related to the engineering approach to estimating the long run average cost of production is to estimate the parameters of a prespecified production function, generally of the Cobb-Douglas or CES type, to test for the existence of increasing returns to scale.* This approach is an indirect test of economies of scale because it explicitly assumes that the industry's production processes conform to simple laws of production and implicitly assumes that all firms pay the same prices for the factors of production, regardless of their size.

Properly applied, this technique involves choosing a cross section of plants of widely different sizes and fitting plant level observations

---

*For a review of these prespecified production functions and the definition of increasing returns to scale see Chapter 1.

on output and the primary inputs (capital, labor, energy and materials)[*] to a prespecified production function through regression analysis. One of the prespecified properties of these functions insures that if a plant's original planned capacity is increased by, for example, 13 percent as a result of a 10 percent increase in the primary inputs, then all plants (regardless of size) will experience the same 13 percent gain in planned capacity for a 10 percent increase in the consumption of primary inputs. This property means that this approach is at best only capable of identifying the returns to scale associated with the average sized plant.[†] Thus, what is most often of greatest interest to economists and managers, that is, the potential for further gains in the efficiency with which resources are used by plants operating on a scale larger than current practice, is unlikely to be uncovered through this approach unless the industry's processes conform closely to the prespecified production models that represent the core of this analytical framework. Furthermore, in the event that an industry's processes are specified accurately by a function of this type, the change in the ratio of one of the plant's inputs to plant output (for example, the inverse of output per man-hour) for increases in scale would be sufficient to indicate the extent of scale economies,[‡] thus obviating the need for a production function as an analytical intermediary. Thus, those who employ the prespecified production function approach in analytical studies of scale tacitly are admitting that they expect, at best, to discover the returns to scale that would apply to the plant of average size, and

---

[*]Output may be measured by sales, and materials by materials cost; or materials may be excluded, and output (net of materials consumed) measured by value added. The latter is the most common approach. Labor is typically measured by the number of employees or man-hours, and capital is measured most commonly by gross book values of the capital stock, since net values typically underestimate the value of the capital stock.

[†]Actually, if increasing returns to scale are discovered through applying this technique, the magnitude of these returns should be associated with a plant of less than average size, because the parameters of the function can be estimated through least-squares only after the production equation has been transformed to a log-linear form—a form that puts disproportionately heavy weight on the goodness of fit for small plant scales.

[‡]For example, if the inverse of output per man-hour fell by 10 percent for a doubling of the scale of plant, this would indicate that long run average cost fell by 10 percent. This conclusion follows from the implicit assumption that all firms, regardless of size, pay the same prices for the factors of production. This means that each plant, regardless of scale, would use the same mix of resources; and thus, any input could be used as a proxy for all inputs.

that they do not expect close conformity between the model and the actual production processes. It is difficult to see what benefits may be derived from this approach other than a crude measure of the likelihood that the average plant size will tend to increase over time because of potential scale economies. [18]

When these pessimistic conclusions are coupled with the serious measurement problems and the paucity of data confronting efforts to employ this technique, [19] it is apparent that even the mediocre findings representing the best possible results this approach has to offer rarely will be achieved.

THE INDUSTRY CASE STUDY APPROACH

Instead of attempting to derive a long run average cost curve that is of only temporary significance and of doubtful meaning[*] or trying to deduce the extent of economies of scale through simple economic models pertaining to scale, the industry case study approach represents an alternative with the potential of deriving greater insight into the role played by scale in the evolution of industrial processes. The essential difference between the methods reviewed above and this approach is that the industry case study approach is based on the premise that all the economic and technical characteristics of the industry have a role in determining both the potential of scale and the diffusion of large scale techniques of production throughout the industry. Each industry is treated as partly unique, based on the premise that no single, standardized approach to studying scale will succeed completely.

Of all the factors that play a role in determining the potential and practice of large scale techniques of production (which include the growth in demand for the industry's products, the structure and conduct of firms in the industry, and changing conditions in supplying industries), perhaps the most important is technological change, since changes in technology can remove obstacles that temporarily can block further gains in efficiency through increases in scale. An excellent example of the intertwined nature of technology and scale can be found in the petroleum refining industry.

In a mid-1950s study, John Enos was able to isolate two phases of development for each of four successive cracking process innovations in petroleum refining from 1913-55.[20] The first phase could be labeled the innovation phase and the second the scale phase.[21] The end of the innovation phase was defined by Enos as the year in which the first commercial plant embodying that technology was put

---

[*]See Chapter 1 for a review of the ambiguity associated with the long run average cost curve.

into operation; and the scale phase ended when "improvements virtually ceased, " which he claimed closely coincided with the construction of the first commercial plant embodying a new process innovation. [22]

Measuring improvements in operating efficiency by the reduction in the ratios of four basic inputs (labor, raw materials, fuel, and capital) to output (measured in ton-miles equivalents), Enos was able to conclude, ". . . the [scale] phase is as significant in its economic effects as is the [innovation]. [23] In fact, his data indicated a greater drop in the input-output ratios during the processes' scale phases than between the ends of the scale phase of one process and the innovation phase of its successor; albeit the reductions that occurred during the scale phases cannot be attributed entirely to increases in scale but are attributable partly to additional technical advances and improvements in operating knowledge.

In short, the Enos study strongly supports the twin hypotheses that scale and technological change are intertwined forces, and that distinctive improvements in the latter can pave the way for further increases in the former. To isolate partially the effects of scale, Enos found it necessary to explore the technological history of the petroleum refining industry in some detail, an approach that is characteristic of industry case studies.

Accompanying the richer potentials of the industry case study approach are analytical problems not normally encountered in the standard "arena" of microeconomic theory. Industry equilibrium must be discussed in a dynamic (rather than static) setting, since the focus of the analysis is on explaining changes in the scale of production. In conventional theory, the focus is centered on the final equilibrium of firms after equilibrium-disturbing changes have been absorbed; very little attention has been paid to the speed with which firms react to such changes. Most of the theoretical literature on equilibrium deals either with the market structure extremes of perfect competition or (somewhat trivially) monopoly; or it deals with the stability of special oligopolistic equilibriums arising from specific strategies employed by the firms. Most of the theoretical work on technological change is found in macroeconomic growth theory. (Some notable exceptions include the research efforts of Bela Gold, Edwin Mansfield, and the late W. E. G. Salter. [24] Little theoretical research has been done on the problems of analyzing the production processes of plants in which multiple products are produced, except for some recent developments in linear activity analysis.

Gold's Model

Most of these problems were treated by Gold in 1955. [25] He was concerned with strengthening the foundations of industrial productivity analysis. To accomplish this task, he dealt successively with the need to modify the basic objectives and concepts of productivity analysis, the measures and analytical models of productivity, and the sources, nature, and effects of productivity adjustments (particularly from the management viewpoint). His basic model was concerned with the dynamic effects of technological change and the evolution of large scale production techniques on factor proportions, unit costs, and ultimately on managerial strategy.

In its simplest form, Gold's model is analogous to the general production function of economic theory but contains none of its inbred shortcomings. It consists of three productivity ratios and three factor proportions, which encompass the direct factors of production: capital, labor, and materials. [26]

These variables are designed to analyze historical developments and meet the basic requirements of practical decision making by management, by covering,

(a) Changes in the level of each category of input requirements per unit of output, including materials, facilities investment and salaried personnel as well as direct labor;

(b) Changes in the proportions in which inputs are combined, both in order to take account of substitutions (e.g., buying more highly fabricated components instead of making them, or replacing labor with machinery) and also in order to differentiate between changes in the productivity of major as over against minor inputs;

(c) Differences between the productivity of inputs when they are fully utilized and when their contributions are reduced by idleness (e.g., as in the case of under-utilized equipment);

(d) Variations in all components of this 'network of productivity relationships' as viewed simultaneously by managers capable of adjusting relationships among them in the interests of improving aggregate performance relative to specified criteria. [27]

A major strength of this system is its ability to investigate production processes with multiple products. Physical output is measured by a price-weighted index of the multiple products, in the manner proposed by Solomon Fabricant. [28] Materials volume can be measured by the same technique. Capital is measured by fixed investment. But instead of relating fixed investment directly to output, it is linked indirectly to output through plant capacity (which is the "product" of the fixed investment expenditures).

Capacity is commonly the most difficult variable to measure in Gold's model, but by adjusting for changes in the important determinants of "practically sustainable capacity" (product mix; hours per shift, shifts per day, and such; the general quality and availability of the variable inputs; and product and factor prices), capacity can be measured with reasonable accuracy in all but the most labor-dominated processes (in which the capital input is likely to be of minimal relative value). Indeed, as Gold indicates, all production scheduling is based on an experimentally determined practical capacity adjusted for any important shifts in product mix.

Gold investigates likely adjustments in the productivity ratios and factor proportions under a variety of process innovations having their major initial impact on one (or more) of the inputs. He demonstrates that the subsequent adjustment patterns depend on both market constraints (well recognized in the literature of theoretical economics) and (less recognized in the literature) technological constraints (such as the extreme cases in which the nature and volume of one input almost completely determines the capacity of the production process and the qualities and quantities of the other inputs). Since any study of changes in scale over time must concern itself with these effects of innovations, some of Gold's observations are valuable.

In the opinion of some, Gold's basic production model may appear to be weak because of its avoidance of detailed analytical prescriptions. But this avoidance is actually one of the model's strengths. In the field of law, the strength of a legislative statute affecting commerce is weakened if its intent is made too explicit because it quickly becomes outdated. It is considered better technique to express only the general intent of the lawmakers in the statutes and leave the details of interpretation to the courts, who may then consider the unique features of each case when rendering a decision. The same reasoning can be applied to models of production and cost in economics.

Salter's Model

Many of the problems encountered in a dynamic analysis of industrial processes were treated also by Salter in 1960.[29] He was concerned with the speed of diffusion of technological innovations and the relationship between productivity and technological change in the sense of how improvements in the former are caused by the latter (the relationship between these two forces is determined partly by the speed of diffusion of new innovations).

Although he considered models of both production and the effects of improvements in production methods on the average performance of all firms in the industry, his basic production model is not sufficiently different from the standard textbook model to warrant review in this study. Hence, attention will be focused instead on his model for the diffusion of technological progress (including increases in the scale of plant operations).

Salter began with a simple model and then gradually relaxed constraints to approach reality. In its basic form, Salter's industry model is an array of indivisible plants that produce a single, homogeneous product at full capacity. The average variable cost of production in each plant depends on the level of technology embodied in the plant; the more developed the technology, the lower the average variable cost. The industry supply curve is an upward moving step function, with the length of each step representing the capacity output of the plant (or plants) embodying the same level of technology and the height of each step representing the average variable cost of production of those plants. Assuming a competitive market for the industry's product, the equilibrium level of output is determined by the intersection of the industry supply curve with the industry demand curve. Whenever technology advances to the point where the present value of revenues in excess of variable cost just exceeds the investment cost of a new plant embodying the latest technology, a new plant is built—representing an increase in supply—and the result is to force the marginal plant (or plants) with the highest average variable cost to shut down. Whenever the industry demand curve increases, the industry price increases, inducing new plant construction until the price is forced back down to its former level. The implications of this simple model are:

(1) the faster the rate of technological progress, the greater the reduction in price, the faster the increase in output, and the faster the shut-down of marginal plants; and
(2) the faster the increase in demand, the greater the installation of new plant capacity embodying the technology of that time period.

The dynamic, equilibrium-determining relationship in this model is:

equilibrium price = best-practice unit total cost
                = unit operating (variable) cost of plants on
                   the margin of obsolescence

Either the addition of new capacity or the scrapping of old capacity causes increases in productivity, giving Salter his basic desired relationship. The refinements in this basic model attempted by Salter were:

(1) allowing for short run adjustments by plants to production levels below or above normal capacity;
(2) recognition that plants are not indivisible and that introduction of new technology on the equipment unit level is possible, except that new equipment may not interrelate as well with existing, older equipment and that the installation cost may be higher in an older plant;
(3) incorporating the external (to the industry) resale value of old equipment in the shut-down condition—revenues must cover operating

costs plus interest on resale values and the decline in the scrap value of the old equipment;

(4) incorporating quality changes in the product due to technological progress; and

(5) dropping the assumption of perfect competition in the output market. In this case, marginal revenue replaces demand in the equilibrium condition, except that the pressure on plants with monopoly power to retain maximum efficiency is now internal (profit maximization) instead of external (threat of bankruptcy).

Criticisms of Salter's model stem from the weaknesses of some of his assumptions and the fact that certain aspects of his model are not practically applicable. The weaker assumptions of his model include the implied homogeneity of products in the industry and the implied insignificance of geographic factors;* lack of recognition of the importance of product differentiation for promotional purposes; the implication that managers are able to identify their demand curves and, hence, marginal revenue curves; and the related implication that profit maximization represents their sole motivation.

Application of Salter's model to the plants in a given industry fails, in general, on the paucity of data pertaining to his key variables. In particular, it may be difficult to identify expansion of industry capacity with increases in scale, improvements in technology, or increases in demand as individual factors. Data on variable cost are difficult to find at the plant level and often do not relate to full capacity operations. The determination of demand curves can be extraordinarily difficult. If the majority of technological innovations are incorporated in existing plants (to lower costs, extend the life of the plant, and possibly expand plant capacity), the roles played by the cost levels of all plants in the market and the level of aggregate demand in regulating the diffusion of innovations throughout the industry is less important than when the technological improvements can be incorporated only through new plant construction. In such cases, Salter's model must be applied at the sub plant (that is, equipment unit) level.

SUMMARY

Caleb Smith's assessment that simplifying assumptions necessary to develop theoretical concepts such as long run average cost block subsequent attempts to verify empirically those concepts certainly seems confirmed in most of the empirical literature on economies of

---

*However, Salter applied his model to British industries, where geographic factors are not as important as they are in the U.S.

scale. We are confronted with another variation of the conflict between theoretical and applied economics, cynically summarized by some as theory without measurement versus measurement without theory.

Yet, it is the feeling of some economists that a careful study of the multitude of factors affecting scale in a given industry—factors either ignored or treated in a hands-off fashion in many empirical studies attempting to investigate large sectors of the economy with minimal effort—will permit more headway to be made in studies of scale.

NOTES

1. C. Smith, "Survey of the Empirical Evidence of Economies of Scale," in Business Concentration and Price Policy, A Conference, sponsored by the National Bureau of Economic Research (Princeton: Princeton University Press, 1955), pp. 213-38.

2. B. Gold, Explorations in Managerial Economics (New York: Basic Books, 1971), p. 3.

3. Smith, op. cit., p. 214.

4. Ibid., p. 215.

5. Smith, op. cit., p. 220.

6. J. Johnston, Statistical Cost Analysis (New York: McGraw-Hill, 1960).

7. C. Smith, op. cit., p. 221. In an earlier study, Joel Dean [Managerial Economics (Englewood Cliffs, N. J.: Prentice-Hall, 1951), p. 299] concluded that four methods of estimating the relationship between long run average cost and plant size were worth considering.

(1) Analysis of changes in actual cost which accompanied the growth of a single plant over time.

(2) Analysis of differences in actual cost of plants of different sizes operated by separate firms and observed at the same time.

(3) Engineering estimates of the alternative cost where the same technology of manufacturing is used in plants of different sizes.

(4) Analysis of differences in actual costs of different sized plants operated by one corporation.

But it is difficult to perceive how the first method could be brought into conformity with the assumptions underlying the long run average cost curve of static microeconomic theory. The second and fourth methods correspond to Smith's statistical approach, and the third method corresponds to Smith's engineering approach.

8. B. Gold, op. cit., p. 181, reprinted from "Economic Effects of Technological Innovations," Management Science 11, no. 1 (September 1964): 107.

9. J. Bain, "Economies of Scale, Concentration, and the Condition of Entry in Twenty Manufacturing Industries," American Economic Review 44, no. 1 (March 1954): 15-39.

10. H. Chenery, "Engineering Production Functions," Quarterly Journal of Economics 63, no. 4 (November 1949): 507-31.

11. Ibid., pp. 507-08.

12. In the areas where the transformation processes can be precisely modeled, Chenery states (Ibid., p. 510), "In order to use [the input combinations found to be inefficient under current prices] conveniently, the economist must abandon his convention of using one dimensional inputs and use multi-dimensional inputs as the engineer does." But such possibilities are limited (Ibid., p. 530).

> In only a relative small proportion of cases do engineers understand well enough what goes on in the transformation of materials into product to reduce the transformation to an analytical model. Yet without such a model, the only productive possibilities which are known precisely are those which have been tried on a commercial scale . . . the only combinations which have any accurate meaning for blast furnaces or turbines are those which have been built.

13. G. Stigler, "The Economies of Scale," The Journal of Law and Economics 1 (October 1958). Stigler credits Willard Thorp with pioneering this approach in a 1924 census monograph.

14. T. R. Saving, "Estimation of Optimum Size of Plant by the Survivor Technique," Quarterly Journal of Economics 75, no. 4 (November 1961): 569-607.

15. Called "establishments" in the parlance of the Census of Manufactures (Washington, D. C.: U. S. Dept. of Commerce, Bureau of the Census, 1963), which is often used as the primary source of data in such studies.

16. T. R. Saving, op. cit.

17. Ibid., p. 579.

18. As an example of a fairly current application of this technique, see John R. Moroney, "Cobb-Douglas Production Functions and Returns to Scale in U.S. Manufacturing Industry," Western Economic Journal 6, no. 1 (December 1967): 39-51. It must be noted with some disbelief that Moroney selected two-digit Standard Industrial Classification Code (SIC) industries (there are only 20 in all manufacturing) as the basis for defining supposedly homogeneous products and selected states, instead of plants, as the basic production units! On the basis of this incredibly misspecified model (which probably was selected because of lack of data on individual plant operations), Moroney heroically concludes, citing "extremely good statistical fits of the regression equations," that his results "are consistent with the hypothesis" that in most industries there are constant returns to scale; that is, "there is a broad range of 'optimal' plant size." Ibid., p. 40.

19. An excellent discussion of the weaknesses of the prespecified production function, the sources of data, and the measurement problems encountered in this approach can be found in Alan Waters, "Production

and Cost Functions: An Econometric Survey,'' Econometrica 21, no. 1 (January 1963): 1-61.

20. John Enos, "Innovation in the Petroleum Refining Industry," in The Rate and Direction of Inventive Activity (Princeton: National Bureau of Economic Research, 1958), pp. 299-321. The four processes were: the Burton, the tube and tank, the Houdry, and fluid catalytic cracking.

21. Ibid., p. 317. Improvements during the scale phase included, "the construction of larger units to take advantage of inherent economies of scale; the adoption of ancillary advances by other industries; and the increase in operating knowhow."

22. Ibid.

23. Ibid., p. 319.

24. Some notable publications of their work include, Bela Gold, Foundations of Productivity Analysis (Pittsburgh: University of Pittsburgh Press, 1955); "Economic Effects of Technological Innovations," Management Science 2, no. 1 (September 1964) pp. 105-34; and "The Framework of Decision for Major Technological Innovation: Values and Research," in Values and the Future, edited by K. Baier and N. Rescher (New York: The Free Press, 1969), pp. 389-430. E. Mansfield, "Technical Change and Rate of Imitation," Econometrica 29, no. 4 (October 1961), pp. 742-66; and "Speed of Response of Firms to New Techniques," Quarterly Journal of Economics 77, no. 2 (May 1963), pp. 290-309. W. E. G. Salter, Productivity and Technical Change (Cambridge, England: Cambridge University Press, 1960).

25. B. Gold, Foundations of Productivity Analysis, op. cit.

26. Ibid., pp. 62-88. The three productivity ratios are:

$$\frac{Output}{Man\text{-}Hour}, \frac{Output}{Materials\ Volume}, \text{ and } \frac{Capacity}{Fixed\ Investment}$$

Capacity, rather than output, is treated as the product of investment expenditures to differentiate between the potential contribution of the capital goods and the extent to which this potential is not realized due to under utilization of capital goods. The factor ratios are:

$$\frac{Materials\ Volume}{Man\text{-}Hours}, \frac{\frac{Output}{Capacity} \times Fixed\ Investment}{Man\text{-}Hours}$$

(that is, the ratio of capital services to labor), and

$$\frac{Materials\ Volume}{\frac{Output}{Capacity} \times Fixed\ Investment}$$

Any one of the productivity ratios may be treated as the resultant of a factor proportion and another productivity ratio, indicating the interaction between these variables. For example,

$$\frac{\text{Output}}{\text{Man-Hours}} = \frac{\dfrac{\text{Output}}{\text{Capacity}} \times \text{Fixed Investment}}{\text{Man-Hours}} \times \frac{\text{Capacity}}{\text{Fixed Investment}}$$

27. B. Gold, "Technology, Productivity and Economic Analysis," Omega 1, no. 1 (1973): 11.

28. B. Gold, Foundations of Productivity Analysis, p. 93. Gold, op. cit. Solomon Fabricant The Output of Manufacturing Industries, 1899-1937 (New York: National Bureau of Economic Research, 1940), p. 24.

29. Salter, op. cit.

3

A PROPOSED MODEL
FOR ANALYZING THE
EFFECTS OF SCALE

THE SCOPE OF THE MODELING PROBLEM

Efforts to describe the relationships between materials, energy, capital, labor, and output at the level of a representative plant in a particular industry by employing general (but prespecified) production functions encounter a number of difficulties. The first set of problems is the result of this particular approach to modeling. It includes the previously discussed limitations of the prespecified production function and crucial problems of measuring the inputs—particularly capital.

A second group of difficulties is related to the concept of the representative plant. The extent of horizontal and vertical integration on plants producing products classified in a certain industry may differ widely. Differences in vertical integration may cause significant variations in the relationship between output (measured as value added) and capital and labor.[*] Variations in the product lines produced by plants in the industry will cause similar variability in the relationship between the inputs and output. In this same context, plant scale is not determined correctly by the output or capacity of the plant. This would be true even if significant variations in vertical integration and product coverage did not exist because of the additional possibility that some larger plants are only larger clusters of facilities employed in smaller plants and because basic production techniques (or the age of the capital stock) may vary among plants.

These problems have been surmounted partially in industry case studies, such as the Enos study of petroleum refining reviewed in the previous chapter. The drawbacks of this approach stem from the

---

[*]Or, the variations may occur between output (measured in physical units) and capital, labor, and materials.

tendency of many such studies to be excessively descriptive and from their failure to cover the operations of all plants in the industry. It would still be desirable in the industry case study approach to retain the mechanism of a production function or some related analytical framework that summarizes the physical-engineering relationships between inputs and output at different scales of plant. It is only by applying some method of isolating the contribution of each input to the level of output, and the combined effect of these individual contributions, that the empirical analysis can probe beneath the simple average cost-scale relationships to discover the basic underlying forces at work.

The relationship between the aggregate, plant level production function and a totally descriptive specification of the production possibilities of the plant indicates the basic nature of the problem of modeling production. The aggregate production function may be treated as a result of applying weights to each class of inputs (for example, "capital," covering machines, land, warehouses, buildings, and other facilities) in the descriptive set of production possibilities and subsequently fitting the weighted inputs to a particular function. The set of production possibilities represented by the aggregate function is considerably smaller than the descriptive set of production possibilities because many of the latter possibilities may be discarded as inefficient once the weights have been applied. For example, if $(K_1, L_1, M_1) <$ (K_2, L_2, M_2), where K, L, M represent the weighted capital, labor, and materials inputs, the second combination may be rejected as inefficient. But in the descriptive set it is likely that each of $(\bar{K}_1, \bar{L}_1, \bar{M}_1)$ is not less than its counterpart in $(\bar{K}_2, \bar{L}_2, \bar{M}_2)$, where $\bar{X}$ (for $X = K$, L, M) represents a vector of real numbers $(x_1, x_2, \ldots, x_n)$, and where $(\bar{K}, \bar{L}, \bar{M})$ is the unweighted counterpart of (K, L, M). In short, weighting schemes represent more information about the nature of the production processes—and, in particular, imply that it is possible to measure accurately capital, labor, and materials with a single real number valued variable.

It is possible to specify a relatively greater number of inputs in the plant level production function by defining each input more narrowly (for example, skilled and unskilled labor instead of hourly employees). The greater the number of inputs specified, the closer the production function can be approximated by a descriptive set of production possibilities. A greater number of input categories may reduce the measurement problem, but this approach complicates the interpretation of the production function and reduces the likelihood of a good fit (judged by some normative criterion of accuracy, for example, R^2 in excess of 0.9). Hence, the problem of modeling production at the plant level also may be viewed as a search for a reasonable tradeoff between input measureability and simplicity in the model's specification, as well as a search for an adequate data base of plant level observations.

THE BASIC PRODUCTION MODEL

Because the primary focus of this study is on the iron and steel industry, the model developed in this chapter is tailored to the particular characteristics of fundamental production processes in that industry, although the model could have wider applicability. First, the process of interest tends to be either capital-dominated (that is, the labor force primarily is engaged in starting, stopping, tending, maintaining, and repairing the capital facilities while materials play a passive role) or both capital- and materials-dominated (that is, the productive capacity of the capital is strongly dependent on the quality of materials). [1] Materials play a particularly important role in this industry in the first few stages of production, where processes are designed to alter the chemical composition and the physical properties of the materials. The effect of these two characteristics on a model of production are (1) the required services of labor are basically a function of the capital stock (that is, the number, size, and technological characteristics of the machines used in each process); therefore, increasing the number of hourly employees beyond the required level will have little or no effect on the capacity of the plant; and (2) the capacity of the capital facilities depends on the chemical and physical characteristics of the materials.

The extent of vertical integration and the range of products differ substantially among plants in the steel industry. The heterogeneity of the capital goods employed in these plants is so great as to defeat all reasonable attempts to measure them in physical units. (The machines differ in function, size, sophistication, the types of materials from which they were constructed, and durability, to name some of the more obvious sources of heterogeneity). In addition, the basic sources of increases in scale are found at a level of aggregation (of capital facilities) smaller than the plant (except for small, nonintegrated plants). These features of plants in the steel industry suggest that attempts to model production processes should be undertaken at a subplant level. For example, if modeling efforts are concentrated at the level of specific processes or activities (for example, smelting, refining iron into basic steel ingots, and hot rolling and cold rolling operations), the result will be to avoid the differences in the scope of production processes and the range of products among plants, to simplify the problem of measuring capital, and to permit analytical efforts to concentrate on explaining the effects of changes in scale. A more detailed exploration of the relationship between plant level models and process level models of production entails recognizing three basic sources of capital heterogeneity at the plant level, which will be called technique heterogeneity, activity heterogeneity, and scale heterogeneity.

Technique Heterogeneity

In conventional production theory it is recognized that there exist during any production period more than one technique of producing a given volume of output. The physical reality of this generalized fact is that materials, capital, and labor can be substituted for each other. In a capital- and materials-dominated industry, these substitution possibilities imply that machines and facilities of one design can be substituted for machines and facilities of another design, purchased materials can be processed to a greater or lesser extent, and thus the labor requirement for a given volume of output can fluctuate considerably. Presumably, if the substitution is made in the direction of more automated machines (more capital), fewer man-hours will be required to operate the machines to achieve some stated output (less labor). A change in the technique of production is viewed in this study as a major change in capital facilities, such as a shift from open hearth steel furnaces to basic oxygen furnaces. In some cases, it may be difficult to distinguish between a major change and a minor change involving design variations or improvements in existing types of facilities. Fortunately, this classification usually is aided by the technical and engineering literature of the industry. One possible guideline is to determine whether the change can be incorporated into the existing capital stock (called disembodied technical progress) or whether it only can be implemented by scrapping the old stock and replacing it with new equipment (called embodied technical progress).

Activity Heterogeneity

This source arises from the breakdown and organization of tasks within the plant. At any given point in time, it is possible to chart the vertical and horizontal sequences of activities that lead to the array of final products that the plant is designed to produce. Presumably, a plant originally is designed and built with one plant wide production technique in mind. Each existing plant also possesses certain degrees of flexibility that permit modifications in the original production technique as technological progress is made. Thus, a major technological improvement (that is, a change in the technique of production in the sense defined above) may be incorporated into existing plants as well as new plants, despite the likelihood that the existing sequence of activities would be altered.*

---

*This is particularly likely if the major improvement is localized in a small sequence of activities in the plant. On the other hand, if

It is possible to identify more activities as finer gradations of differences are considered. The basic guideline is to identify as different activities those tasks using different forms of physical capital and producing different kinds of intermediate or final products. One specific guideline that might be helpful in plants not dominated by continuous processes is to define an operation as a separate activity if its output can be inventoried at least temporarily. This would include transportation activities within the plant, where the basic operation— a change in position—can be maintained for at least a brief period of time. In cases where the same basic transformation of materials and intermediate products to a more finished state is achieved by two or more different techniques of production in the same plant, that is, the same task is performed by a variety of designs of physical capital, then this basic activity (this specific transformation) can be decomposed into a series of horizontally integrated activities—one activity for each specific design of physical capital—occupying the same position on the spectrum of vertically integrated activities. Once the overall plant operations have been decomposed into this activity framework, the plant's capital stock has been separated into groups that are mutually noncomparable due to fundamental differences in design and/or basic task orientation.

Scale Heterogeneity

Within each activity as defined above, there is often a possibility of varying the capacity of individual equipment units. This possibility represents the fundamental source of increases in scale in the iron and steel industry, as well as in other capital-dominated industries engaged in extractive processes or employing equipment units whose volume or speed of operation is a partial determinant of capacity. Those industrial processes where the relationship between capacity and capital investment is susceptible to estimation by models of the six-tenths rule fall into this category. The difficulty raised by scale heterogeneity is the generally nonlinear relationship between machines of the same basic design and task orientation when they are built in different size configurations. This may be stated another way: a machine of design X with a capacity of 2Q is not twice as much physical capital as two machines, each of design X and with a capacity of Q.

the change embraces practically all activities in the plant, it is doubtful that it could be incorporated economically in an existing plant.

Adjustments for Capital Heterogeneity

If these three types of capital heterogeneity could be eliminated, it would still not be possible to measure capital accurately by enumerating the number of machines because of the multitude of design variations relating to the durability, flexibility, age, precision, dimensions, and so on of the equipment. Nevertheless, eliminating the first two sources of heterogeneity (technique and activity) greatly reduces the problem of measuring capital since the analysis is then focused on a particular activity employing equipment units of a certain basic design, each with a similar task orientation.

Removing technique heterogeneity generally is necessary in studies of scale. Furthermore, most empirical studies of production are not able to incorporate all techniques of production in the model because many of the techniques (that cannot be discarded on the basis of physical inefficiency)[*] currently are unused (and may never have been used) by plants because they are economically unattractive. Hence, most production models are, at best, only capable of explaining "what happened," given past prices, or "what is," given current prices, and are not capable of explaining "what could have happened" or "what could be."

Adjusting for activity heterogeneity by concentrating the analysis of production at the activity level (that is, estimating a production relationship for each identifiable activity) has the advantages previously cited. This is the fundamental level at which capital, labor, and materials are organized. (In some industries, it is likely that some firms entered the industry by concentrating initially on a particular activity and then expanding to include other activities in the scope of their operations.) In cases where changes in scale or changes in production techniques involve higher levels of aggregation, a sequence of activities encompassing the entire change can be selected for analysis. This has the advantage not only of measuring changes in capital and labor requirements but also of indicating the effect of the change on the organization of tasks in the plant. [2] If the reduction in unit capital and labor cost is only slight while the reorganization of tasks is major, this may lead to slow rates of adoption of the change by firms in the industry.

For each activity, the direct inputs are:

(1) materials—either intermediate products of preceding activities within the plant and other plants of the same firm or products purchased from other firms;

---

[*]If $(K_1, L_1, M_1) < (K_2, L_2, M_2)$, where $Q_O = f(K_1, L_1, M_1) = f(K_2, L_2, M_2)$, then the second set would be discarded as physically inefficient.

(2) labor—both salaried personnel and wage earners employed directly in the activity;

(3) equipment and facilities—the physical capital of the activity; and

(4) energy—the input necessary to supply power to such items as machinery (separated from materials for a reason that will be discussed later).

Describing the production relationship in an activity involves describing the relationship between these direct inputs and the activity's output, byproducts, and waste. The activities are interconnected because the intermediate output of one activity appears as a materials input to a succeeding activity and because subgroups of activities share common facilities, such as energy sources, warehouses, land, building, and management, which can be labeled the overhead capital of the plant.*

The degrees of freedom (that is, the generality) of the basic production model at the activity level are curtailed when comparison is made to the general production function of conventional economic theory. Much variability, however, remains. It is still possible to vary the proportions of capital and labor by varying the scale of machines (together with the minor design variations constituting the other sources of capital heterogeneity). It also may be possible to substitute capital for energy by designing the equipment to be more efficient in the engineering sense. But more important, it may be possible to vary the proportions of capital and energy by varying the scale of machines.

Substitutions can be made among some of the materials purchased from other firms or transferred from other plants. This changes the "work" done by the activity and, hence, alters the relationship between capital, labor, and output.† Substitutions also can be made for at least some of the materials produced in other activities in the plant. This affects not only the work done by the activity but also the preceding activities, and it therefore must be considered in a broader framework. [3]

The most general specification of the production function at the activity level is similar to the standard version:

---

*Reductions in the unit cost of these shared facilities and management for increases in the size of plant represent economies of scale that must be estimated at levels of aggregation greater than the activity level.

†The work to be done is determined by the qualitative nature of the materials and the qualitative nature of the activity's output, byproducts, and waste. This describes the precise transformation that must be effected. The "variable" inputs (that is, variable ex ante) are the different combinations of equipment, labor, and energy that can be utilized to effect this specific transformation.

$$Q = F(\bar{K}, \bar{L}; \bar{M}, \bar{E}) \hspace{4cm} (3.1)$$

where any symbol with a bar above it (for example, $\bar{K}$) represents a vector of real numbers and where, more specifically,

(1) $\bar{K}$ represents the number of equipment units and the scale of each unit.

(2) $\bar{L}$ is the volume of each distinct type of labor skill needed to operate and maintain the equipment employed in the activity.

(3) $\bar{M}$ represents the volume of each of the different types of materials (including both complementary materials, which specify the nature of the activity's transformation, and substitute materials, whose inclusion facilitates adjustments for changes in quality that will affect the capacity of the activity and the work done by the activity) that must be considered in the range of practical materials substitutions. Some of the potential materials substitutes included in the design of $\bar{M}$ may be zero valued in $\bar{M}$.

(4) $\bar{E}$ represents the volume of each of the different types of energy inputs that are used in the activity. (Only complementary types are included, for example, electricity and fossil fuel. Where substitution possibilities exist, as fuel oil for natural gas, it is assumed that fuel oil can be expressed readily in natural gas equivalents).

(5) F is a functional relationship that determines the level of output attainable from various combinations of these four inputs. It determines the capacity of a given array of equipment units and labor force for a given mix of materials when $\bar{M}$ and $\bar{E}$ are maintained at their practical, sustainable rate. (A semicolon is inserted between "$\bar{K}$, $\bar{L}$" and "$\bar{M}$, $\bar{E}$" to note the dependence of Q on the flow of both capital and labor services as well as on materials, energy, and the mix of materials.)

In the present context, this general production function can be disaggregated to recognize that we are dealing with capital-dominated processes at the activity level. Thus, labor is not a substitute for capital:

$$\bar{L} = \bar{G}(K; M, E) \hspace{4cm} (3.2)$$

where $\bar{G}$ is an array of functional relationships that specify the level of each labor skill needed to operate the equipment units, given the mixes and throughputs of materials and energy. Furthermore, given that the basic design of the equipment has been determined, the consumption of energy is determined primarily by the number of equipment units, the scale of each unit, and the mix and throughput of materials. That is,

$$\bar{E} = \bar{h}(\bar{K}; \bar{M}) \hspace{4cm} (3.3)$$

where $\bar{h}$ is an array of functional relationships that specify the volume of each type of energy input required to operate the equipment at a rate sufficient to process $\bar{M}$. The mix of energy inputs is expressed as functionally determined by the number and scale of equipment units (given the mix of materials), rather than as directly proportional to output, to allow for the possibility that the scale of equipment units affects the level of required energy.

Substituting equation 3.3 into equation 3.2, we obtain

$$\bar{L} = \bar{g}(\bar{K}; \bar{M}) \qquad (3.4)$$

Substituting equations 3.3 and 3.4 into equation 3.1, the result is

$$Q = f(\bar{K}; \bar{M}) \qquad (3.5)$$

Equations 3.3, 3.4, and 3.5 constitute the basic production model for capital-dominated processes at the activity level.

The mix as well as the throughput of materials is included as a parameter in these three equations to allow for the distinct possibility that materials can affect normal operating capacity through partial or complete substitution of higher quality materials for lower quality materials. (Recall that the vector $\bar{M}$ was defined to include a space for the volume of each material potentially usable in the process; and for some mixes, such as very high or very low quality mixes, some elements of $\bar{M}$ may be zero valued.) The flexibility in the relationship between the design of the equipment and the physical and chemical characteristics of the materials can range from slight to substantial. If there is only slight flexibility, then the materials variable can be dropped from the model's three equations* if they are treated as specifying the relationships between the other inputs and output only at capacity. If the flexibility in materials substitution is stronger, then, for any particular materials mix, the materials variable can be excluded when treating the model's three equations as capacity relationships only.

In these two cases, the three-equation model—now devoid of a materials variable—can be extended to include the following three equations:

---

*There is always the possibility that the quality of materials is not only inflexible due to the equipment design, but that it must of necessity change (most likely, increase) as the scale of operations increases. In this special case, the cost advantage of larger scale operations should be calculated net of the extra cost of the materials. This change must be handled in terms of cost because with fixed proportions it is not possible to apportion the changes in output, relative to the volume of other inputs, between scale and materials.

$$\overline{M} = \overline{m} \cdot Q \qquad\qquad (3.6)$$

$$\overline{B} = \overline{b} \cdot Q \qquad\qquad (3.7)$$

$$\overline{W} = \overline{w} \cdot Q \qquad\qquad (3.8)$$

where $\overline{B}$ and $\overline{W}$ represent, respectively, the byproducts and waste materials generated in the process of producing at capacity (that is, "Q") and were $\overline{m}$, $\overline{b}$, and $\overline{w}$ are technologically determined vectors of constants representing the fixed input-output and output-output relationships that apply at capacity.*

CHANGES IN SCALE AND THE BASIC PRODUCTION MODEL

When the qualities of the materials, output, byproducts, and waste materials are fixed, the work done by the activity is fixed; and attention can be concentrated on changes in capital, labor, and energy in describing the effects of increases in the scale of production. This would be the case when the mix of materials is fixed due to the technical design of the equipment units and is independent of their scale, or when some flexibility in the materials mix is present but the mix is relatively insensitive to changes in materials prices.

Several other possibilities exist, however One arises when the mix of materials is fixed not only by the technical design of the equipment units but also by their scale (because their scale interacts with their design). In this case, a movement toward increased scale would alter the quality of materials and, hence, the work done by the activity.

Another possibility is part of the broader effects of increases in scale when increasing returns to scale are present. In general, while it may be hypothesized that doubling the number of equipment units (given their scale) merely would double all inputs and outputs for full capacity operations in most manufacturing processes† (except for the overhead capital and labor of the plant), increasing the scale of these units may permit some or all of the following savings in inputs:

(1) a reduction in the unit capacity cost of capital facilities, measured as a reduction in fixed investment/capacity;

---

*Equations 3.6, 3.7, and 3.8 could be used in conjunction with equations 3.3, 3.4, and 3.5 at all levels of capacity utilization if the efficiency of materials usage were not affected thereby. In general, however, it seems likely that the unit materials requirements, $\overline{m}$, and the waste materials generated per unit of output would change in response to variations in capacity utilization.

†This is not necessarily true in all processes. Savings in labor are possible for maintenance and repairs operations and other support operations such as materials handling and quality control.

(2) a reduction in the unit labor requirement (direct labor/output at capacity), but because of the capital-dominated nature of the process, it is unlikely that the labor requirement would rise in proportion to the scale of the equipment;

(3) a reduction in the unit materials requirement ($\bar{M}$) by lowering the amount of materials wasted ($\bar{W}$); and

(4) a reduction in the unit energy requirement (energy/output at capacity). This is more likely to occur in extractive processes where energy supplies heat as well as motivates power, because the proportion of the energy input absorbed by heat losses may be reduced.

A change in $\bar{m}$ due to a reduction in waste materials is not a decrease in the work done by the activity and, thus, should be differentiated from a change in $\bar{m}$ due to a change in the quality of materials. The former is a benefit of increased scale, and the latter is either a requirement for increased scale or a cost-reducing decision.

THE EFFECTS OF OTHER CHANGES ON THE BASIC PRODUCTION MODEL

Over the course of time, the qualities of the materials, output, by-products, and waste of an activity tend to change in response to changes in the supply of inputs, in the demand for finished products and by-products to whose production the activity contributes, and to techno-logical change either within the activity (altering some basic features of the activity's equipment) or in preceding and succeeding activities (altering the nature of the activity's materials inputs or output).

To assess the effect of such changes on the production relation-ships of the basic model, consider the following two tables. Table 3.1 classifies the four input variables and the three output variables into primary and secondary decision variables, that is, the variables that management may wish to change qualitatively for a number of reasons. Turning first to the input variables, materials and capital have been classified as primary decision variables because in capital-dominated processes changes in these variables underlie changes in the labor and energy inputs (see equations 3.3 and 3.4 above); that is, the mix of labor skills and the types of energy inputs are determined by the types of materials and the design of the equipment. On the output side, the activity's product has been classified as primary and its by-products and waste as secondary decision variables on the assumption (weakened by the current national concern with the environment) that management is concerned primarily with the qualitative characteristics of the activity's product. Of course, there are occasions when manage-ment seeks to alter the secondary decision variables, but this cannot be accomplished in the present framework without altering some of the primary decision variables.

TABLE 3.1

Decision Variables Subject to Change

Type	Input Variables	Output Variables
Primary	$\bar{M}$, $\bar{K}$	Q
Secondary	$\bar{L}$, $\bar{E}$	$\bar{B}$, $\bar{W}$

Source: Compiled by author.

TABLE 3.2

Structural Variables of the Basic Production Model

Type	Input-Input	Input-Output	Output-Output
Proportionate (equations 3.6, 3.7 and 3.8)	—	$\bar{m}$	$\bar{b}$, $\bar{w}$
Functional (equations 3.3, 3.4 and 3.5)	$\bar{g}$, $\bar{h}$	f	—

Source: Compiled by author.

Table 3.2 classifies the structural variables of the basic production model represented by equations 3.1 through 3.6. Some or all of these structural variables will be affected by changes in the quality of the decision variables.

Table 3.3 considers the first order effect of changes in the primary decision variables on the structural variables of the system. Although there are many intensities of cause and effect, three basic effects of such changes on the nature of the activity may be discerned: no change in the work done by the activity, a change in the work done by the activity (called a change in its scope in Table 3.3), and a loss of the activity's identity caused by a major technological change in the design of the plant's production processes and, hence, its sequence of activities.

When the latter two changes occur, an estimate of the cost saving achieved by altering the primary decision variables often may be made by choosing the smallest sequence of activities in which the scope is the same before and after the change (that is, by including the minimum number of preceding and succeeding activities that form an activity sequence for which the materials inputs and output are the same before and after the change).

TABLE 3.3

Effect of Changes in the Primary Decision Variables on the
Structural Variables of the Basic Model

Altered Decision Variable	If the effect on the activity Is	Then the following structural variables May Be affected
$\bar{M}$	Its scope is altered	Proportionate
Q	Its scope is altered	Proportionate and functional
$\bar{K}$	No change	Functional
$\bar{K}$	Its scope is altered	Proportionate and functional
$\bar{K}$	Its identity is lost and consolidated with other activities or split into two or more activities	Proportionate and functional variables affected in a sequence of activities encompassing the activity being considered

Source: Compiled by author.

It should be emphasized that Table 3.3 reports only the first order effects of changes in the primary decision variables. It is likely there are also second order effects. For example, a change in the quality of materials is reported in Table 3.3 to affect only the proportionate structural variables (because materials are included as arguments in the functional variables). But if such changes cause changes in the quality of the activity's output—which is possible—the functional structural variables also will be affected.

THE BASIC PRODUCTION MODEL AND THE
BLAST FURNACE SECTOR OF THE STEEL INDUSTRY

In the remainder of this study, this model is applied to the blast furnace sector of the steel industry in the 1900-70 period. All of the basic relationships developed in this model are necessary to the study of changes in scale in the blast furnace sector.

(1) The capacity of the blast furnace sector and the total labor input are determined by the number and scale of blast furnaces and the quality of the iron-bearing materials.

(2) The energy input is determined strongly by the quality of materials and less strongly by the scale of blast furnaces.

(3) The unit labor and energy inputs are also determined in part by the rate of utilization and the design of blast furnaces.

(4) The proportion of materials wasted changed considerably during the period as did their quality. Quality was particularly changed during the last two decades, significantly altering the work done by the blast furnace sector.

(5) Product quality improved somewhat during this period.

Chapter 4 provides a brief overview of the processes and equipment used in the blast furnace industry. Chapter 5 details the changes in materials quality and waste, that is, explores changes in $\bar{m}$ and $\bar{w}$ where $\bar{M} = \bar{m} \cdot Q$ and $\bar{W} = \bar{w} \cdot Q$. Chapter 6 develops the relationship between energy, equipment, and materials quality, that is, estimates the parameters of $E = \bar{h}(\bar{K}; \bar{M})$. Chapter 7 explores the functional relationship between output (at capacity and below capacity), equipment, and materials quality $[Q = \bar{f}(\bar{K}; \bar{M})]$; and Chapter 8 reports the changes in the labor input.

NOTES

1. Bela Gold, Foundations of Productivity Analysis (Pittsburgh: University of Pittsburgh Press, 1955), pp. 189-91.

2. A more detailed discussion of the effect of major technological change on the plant's sequence of activities is presented later in this chapter.

3. The broader framework is discussed later in this chapter.

CHAPTER

4

THE BLAST
FURNACE INDUSTRY

THE BLAST FURNACE PROCESS AS A STEP IN STEELMAKING

In the steel industry, a completely integrated sequence of activities includes mining operations, materials preparation, materials transportation, additional materials preparation, smelting iron-bearing materials in blast furnaces, refining pig iron in steel furnaces, rolling steel ingots into basic shapes, and further shaping and heat-treatment processes.

Iron ore, coal, and limestone (flux) are mined in large quantities, processed to some extent (basically crushed, cleansed of impurities, and graded according to key characteristics), and shipped to blast furnace plants where they are prepared for consumption in blast furnaces. Coal is processed in byproduct ovens to produce coke. [1]

From 1.5 to 2.0 tons of iron-bearing materials, 0.6 to 1.1 tons of coke, and 0.2 to 0.5 ton of limestone were consumed per ton of iron produced during this century, in addition to large quantities of air and water. Thus, anywhere from 2.4 to 3.6 tons of solid materials were handled per ton of iron produced. (The higher figure is applicable to the early part of this century, and the lower figure is applicable to recent years.)

These materials are charged into the tops of blast furnaces and preheated air is forced (blown) through them, which ignites the coke, reduces the iron oxides to nearly pure iron, and progressively melts the materials as they descend to the bottom of the furnace. At intervals of four to six hours, the furnaces are tapped and the molten iron flows into ladle cars that transport it to steel furnaces for further refining or, occasionally, to a casthouse where it is cast into pigs (iron ingots).

The iron products can be broken down into subgroups according to product use or difficulty of production. In terms of production difficulty only ferroalloys (as distinct from pig iron) are significantly more costly to make (both in terms of ores consumed and other inputs consumed).

Since ferroalloys comprised only 2 to 3 percent of total output in the 1900-70 period, they are an inconsequential part of this study. (Ferroalloy production was deleted from this study wherever possible.)

In terms of product use, pig iron products can be classified as either direct inputs to steelmaking departments or semifinished goods earmarked for use in the production of finished goods constructed of iron. Basic and Bessemer iron are transported immediately to steelmaking departments; foundry, malleable, and other minor categories of iron (including ferroalloys) are cast into pigs. Basic and Bessemer iron comprised 64 percent of total iron output in 900, 77 percent in 1920, 88 percent in 1940, and 90 percent in 1960.* Thus, blast furnaces became more integrated with steel plants over time.

Since this study encompasses all iron products except ferroalloys, the latter grouping is significant only insofar as it affects (1) the labor requirement in blast furnace plants because casting requires additional labor and (2) the growth in the size of some blast furnaces over time because, in certain regions, the demand for foundry and malleable iron and ferroalloys was limited.

THE BLAST FURNACE AND ITS ANCILLARY EQUIPMENT

Figure 4.1 presents the salient proportions of a modern blast furnace. The furnace has an outer shell of steel and an inner shell of fire-resistant brick. The outer shell is equipped with baffle plates to retard the descent of the water coolant. The inner diameter of the furnace shaft increases from the top to the bosh area to promote the descent of the materials, which are still in a solid or semisolid state until they reach this area. From the bosh area to the hearth area, the inner diameter of the shaft decreases as the melting materials contract in volume, filling the interstices (or voidage) between the solid materials.

The difference between the hearth and bosh diameters and the distance between the bosh area and hearth area are dictated by the size of the particles of materials consumed in the furnace. If coarse materials are used, the contraction of the melting materials is greater; and the bosh area is lowered and widened to accommodate this contraction. In this study, the hearth diameter is used as a measure of blast furnace size. † This is a justifiable measure of size because, as new furnaces expanded in size from 1900-70, increases in the hearth diameter,

---

* Within this two-product grouping, Bessemer iron comprised 88 percent in 1900, 42 percent in 1920, 18 percent in 1940, and 5 percent in 1960.

† The bosh diameter is used when data on the hearth diameter are not available.

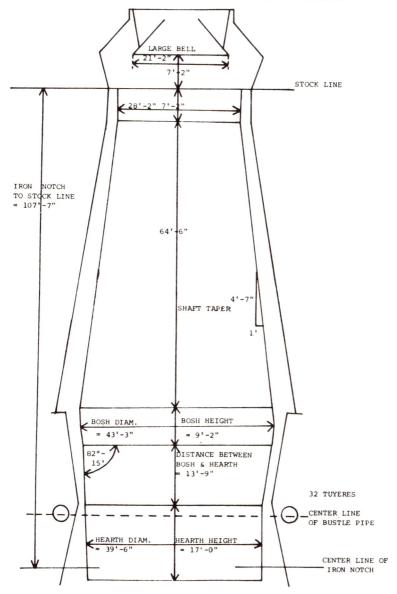

FIGURE 4.1
A Proposed Japanese Blast Furnace with
Dimensions Similar to Modern U.S. Furnaces

LARGE BELL
21'-2"
7'-2"

STOCK LINE

28'-2" 7'-2"

IRON NOTCH
TO STOCK LINE
= 107'-7"

64'-6"

4'-7"
SHAFT TAPER

1'

BOSH DIAM.   BOSH HEIGHT
= 43'-3"   = 9'-2"

82°-
15'   DISTANCE BETWEEN
                 BOSH & HEARTH
                 = 13'-9"

32 TUYERES

CENTER LINE
OF BUSTLE PIPE

HEARTH DIAM.   HEARTH HEIGHT
= 39'-6"   = 17'-0"

CENTER LINE OF
IRON NOTCH

47

bosh diameter, and other diameters of the furnace shaft accounted for more than 90 percent and increases in height for less than 10 percent of the increase in the volume of the furnace.

The top of the furnace generally has three openings. Two of them (located to the side of the large bell, 180 degrees apart) are called downcomers. They are pipes through which the escaping gases are guided to gas cleaning devices, which remove the particles of ore and coke from the gases so the gases can be used to heat the furnace's stoves and power the blowing engines, while the ore and coke can be recycled through the furnace. The third opening is blocked by the large bell and a smaller bell above it. When materials are charged into the furnace the large bell is closed, and the small bell is opened. Then the small bell is closed, and the large bell is opened, allowing the materials to fall into the furnace without losing blast furnace gas.

The materials are carried to the furnace top through the skip hoist – buckets mounted on a belt that runs diagonally from the stock house (where moderate amounts of iron-bearing materials, coke, and limestone are stored) to the furnace top and back again.

The hot blast stoves, either three or four in number, are located in a straight line with the blast furnace. Generally, each has a volumetric capacity equivalent to that of the blast furnace. They are cylindrical structures, built with steel outer shells and filled with a checker-work of firebrick. The heating capacity of the stoves usually is measured by the surface area of the firebrick. The blast furnace gas is burned at the bottom of the furnace until the firebrick becomes very hot. Then air is blown by a turboblower through one stove at a time, where it is heated to anywhere from 800 degrees Fahrenheit (in the early 1900s) to 2200 degrees Fahrenheit (in the best-practice operations today); and then it passes into the blast furnace through openings called tuyeres. While one stove is being blown, two or three others are being heated so the hot air blast is maintained on a continuous basis.

Major improvements in blast furnace design have occurred in the following areas:[2]

(1) Improvements in bell design have led to improvements in the distribution of solid materials and increases in the pressure within the furnace and, thus, have increased production rates and decreased operating irregularities.

(2) The firebrick lining was made more durable by the adoption of power processing in the 1920s and vacuum processing in the 1930s, increasing the length of time the furnace could be operated between relines (called the campaign time).

(3) Blowing engines were improved. The first gas-powered blowing engines were installed in 1903 and rapidly were adopted by blast furnace plants. In 1910, the first turboblowers were installed and became standard operating practice in the industry by the early 1920s.

(4) Materials handling was mechanized, replacing men, shovels, and wheelbarrows. The mechanization of the charging operation, almost

complete by 1920, removed a bottleneck that was threatening to curtail further growth in the size of new furnaces. Thus, by 1920 the blast furnace sector was clearly capital-dominated, although it was basically capital-dominated even when materials handling was labor intensive.

(5) Gas cleaning operations were improved by new equipment. In the early part of this century, various washing techniques were employed to remove a large proportion of the smaller particles. After the initial adoption of the Cottrell electrostatic precipitator in 1919, there were significant improvements during the next decade in the purity of blast furnace gas. The improved purity of this gas paved the way for large increases in the surface area of firebrick in a stove of given dimensions, by allowing the size of the checker openings to be reduced greatly.

(6) A number of control devices were devised. For example, the stockline recorder allowed operators to monitor the rate at which materials descended, giving them a forewarning of many operating irregularities. Temperature, pressure, and humidity gauges were developed, giving operators better control over the reactions within the furnace. During the last decade, automated controls have been installed on some furnaces to eliminate delays caused by the limited ability of human operators to react to adverse changes within the furnace.

(7) Labor intensive casting operations were replaced by mechanical casting machines. This substitution was generally complete by the early 1920s.

NOTES

1. U.S. Bureau of Mines, Minerals Yearbook (Washington, D.C.: U.S. Department of the Interior, 1915, 1920, 1925, 1930). In 1900, only 5 percent of U.S. metallurgical coke production was produced in byproduct ovens; the other 95 percent was made in beehive ovens, located primarily at coal mines. Thereafter, byproduct coke production grew rapidly, reaching 34 percent in 1915, 60 percent in 1920, 78 percent in 1925, and 94 percent in 1930.

2. Detailed in William T. Hogan, S.J., Productivity in the Blast Furnace and Open Hearth Segments of the Steel Industry: 1920-1946 (New York: Fordham University Press, 1950), pp. 34-38.

This chapter will review the types, absolute and relative amounts, and the iron content of the various iron-bearing materials used in blast furnaces from 1900 to 1970. These characteristics are important determinants of blast furnace performance and the volume of flux and energy materials consumed. *

NOTABLE CHARACTERISTICS

Chemical Composition

The iron-bearing materials are comprised chemically of iron, oxygen (bonded to the iron), water, and various impurities called gangue (most of which is silica, alumina, and manganese).[1]

The iron content is an important characteristic because it determines the maximum possible volume of pig iron production from a given amount of iron-bearing materials. It is determined by dividing the weight of pure iron in these materials by their total weight (including or excluding the water content) and is expressed as a percentage value. When the water content is included in the total weight, this value is called the natural iron content. When the water content is excluded, it is called the dry iron content. Both of these measures are used widely in the industry.

The oxygen content of an iron-bearing material (except scrap) is determined basically by its iron content and the kind of ore from which the material was derived.† The two most common ores consumed by

---

*Their effect on flux and energy consumption will be discussed in Chapter 6, and their effect on blast furnace performance will be treated in Chapter 7.

†Scrap generally has a smaller ratio of oxygen to iron than other materials because it is only partly oxidized.

domestic blast furnaces since 1900 have been hematite (Fe_2O_3) and magnetite (Fe_3O_4), containing 30 percent and 27.7 percent oxygen by weight, respectively. Thus, an iron-bearing material derived from hematite with a dry iron content of 56 percent has an oxygen content of approximately 24 percent;[*] the remaining 20 percent represents impurities.

Because of the relatively narrow range of variation in the ratio of oxygen to iron, once the dry iron content of an iron-bearing material is known, the impurity content of the dry ore can be calculated with reasonable accuracy, although the internal composition of the impurities may vary considerably. Given the internal composition of the impurities, the chemical quality of the iron-bearing materials is higher with a lower water content and a lower impurity content. Stated another way, the natural iron content would be closely indicative of chemical quality if the internal composition of impurities were constant and if the water content remained in constant proportion to the impurity content.

Generally, water is less costly to remove than the impurities in the iron-bearing materials. Thus, if two sources produced iron-bearing materials with the same natural iron content, the material with the higher water content would be considered superior, ceteris paribus.

Physical Quality

The hardness and uniformity of size of individual particles of the iron-bearing materials determine their physical quality; the greater the uniformity and hardness (these are related), the higher the physical quality.

CLASSIFYING IRON-BEARING MATERIALS TO MEASURE CHANGES IN QUALITY

If the iron-bearing materials are classified by basic type, the potential variance in chemical quality that can occur with a given natural iron content will have been reduced greatly within each category in relation to all iron materials. That is, within each basic group, the natural (or dry) iron content will be a more accurate indicator of the chemical quality of the material. In addition, some major variance in physical quality will have been removed.

---

[*]The 24 percent value is only an approximate measure of the oxygen content because certain iron-bearing materials have undergone heat treatment in the presence of reducing and fluxing agents, which can alter (generally lower) the oxygen content.

The iron-bearing materials generally can be classified as treated or untreated ores or scrap. In the jargon of the steel industry, "scrap" has a precise connotation as a material that is almost pure iron. In this study, use will be made of Webster's broader definition of scrap: "manu-factured articles or parts rejected or discarded and useful only as material for reprocessing."[2] Scrap may be generated in the blast furnace plant in pouring or casting operations, in which case it is very high in iron con-tent; or it may be a byproduct of other activities in the steelmaking process, typically mill cinder, roll scale, and open hearth slag.

Untreated ores are simply ores smelted in their natural state (al-though the ores may be blended). Treated ores have been upgraded, or beneficiated, prior to smelting. Essentially, the beneficiation of ores involves a variety of processes designed to alter the physical and chemical properties of natural ores to improve the efficiency of subse-quent processes for producing pig iron.* Beneficiation may be under-taken at ore mines, in blast furnace plants, or both. Typically, concentration—the removal of impurities in the ore—is part of the mining activity. The simplest concentration processes include basic crushing, screening, and washing operations. If this is the extent of the treatment prior to smelting, the resulting iron-bearing materials are called ore concentrates. But the beneficiation process may also include agglomeration.† Agglomeration is the process of combining very fine particles of ore and other materials into larger pieces (one-half inch in diameter). The two most important agglomeration processes are sinter-ing and pelletizing.

Thus, as an approximate method of grouping the iron-bearing materials by key differences in the physical and chemical dimensions of quality, they may be classified as scrap, ores, sinter, and pellets. Scrap is characterized by relatively large, irregular-shaped pieces. Its average iron content has tended to change gradually over time as new technical processes affecting the creation of scrap materials have diffused throughout the industry.

Ores have varied widely in physical nature, from fines to lumps. (Lump ores result in superior blast furnace performance, given the natural iron content.) Ores smelted in their natural state usually are comprised of a mixture of these sizes. But if the ores are beneficiated

---

*Accordingly, beneficiation may be regarded as a partial substitute for smelting operations. The extent to which iron ore is beneficiated de-pends on economic and related technical factors associated with the pro-duction of pig iron and steel products. For example, it has been techni-cally possible for a number of decades (but commercially feasible only recently and only in special situations) to bypass the blast furnace process completely through various intensive beneficiation techniques known as direct reduction processes.

†Today, ores generally are concentrated prior to or as part of agglomeration, although this was not always the case.

through simple concentration processes, it usually is possible to segregate the ore concentrates into various categories on the basis of size and ship the smallest sizes (fines) to sinter plants for agglomeration. Ores and ore concentrates also can vary significantly in iron content, although the average natural iron content of all such ores consumed in the United States changed little from the early 1900s to the mid-1950s, when foreign ores began to supplant domestic supplies in large quantities.

Sinter has been more homogeneous than ores in both size and iron content (but still quite heterogeneous over time); pellets are strongly homogeneous in both respects. On the average, agglomerates have been higher in iron content than ores and—during the last two decades—scrap has had a higher iron content than agglomerates.

Employment of these four categories and the natural iron content of the iron-bearing materials comprises the major effort to control for differences in quality in this study. Two factors tend to change the quality of these materials over time: the quality of ores in their natural state and the extent and effectiveness of the beneficiation processes. Substitution of sinter and pellets for ores and ore concentrates is handled adequately by this framework, but improvements within each category (aside from increases in the natural iron content) are not covered.

The quality of ores in their natural state can be altered significantly if the sources of supply change in relative magnitude. To give a rough indication of changes in the quality of ores in their natural state, the following sources will be considered: Mesabi, other mines in the Lake Superior region, other domestic mines, Canadian mines, and other foreign mines. The quality of ores can also be affected when ore concentrates are substituted for untreated ores, although the major effect of concentration is to raise the iron content and to lower the impurity content of the ores. Nevertheless, a brief consideration of the increased practice of concentration will be presented.

Sinter

Within the category "sinter," the most dramatic improvements were accomplished. Sintering was practiced first in 1910 as a means of utilizing the fine particles of ore and coke, called flue dust, that were blown out of the top of blast furnaces. Thus, early sintering facilities were located in blast furnace plants. The early sintering process entailed combining flue dust, coke breeze (small particles of coke that are a byproduct of coke production), and small amounts of other materials on a grate, heating them until the coke ignited, and fusing the materials into a hard substance with a ceramiclike glaze. Subsequently, the pieces of crude sinter were crushed to reduce them to acceptable proportions. Early sinter production had poor physical characteristics.

It was crushed easily when handled and was thus heterogeneous in size, and it lacked the porosity of natural and concentrated ores. Prior to World War II, blast furnace managers considered sinter to be inferior to natural ores, but a cheap way to recycle the otherwise useless flue dust.

During the last two decades, sinter quality has been improved gradually by increasing the proportions of limestone, roll scale, and sinter fines (fine particles of natural and concentrated ores shipped from ore mines) in the sinter mix. Sinter fines were increased from virtually nothing in 1910 to 50 percent in 1950 and 80 percent in 1960, while the proportion of flue dust was decreased accordingly (see Table 5.6). Today, it technically is possible to produce sinter that is far superior to natural and concentrated ores in terms of chemical and physical properties. High quality sinter is characterized by a high iron content, a low level of impurities, a sufficient quantity of lime* fused to the other materials to reduce greatly or remove the need for charging limestone separately into blast furnaces, adequate porosity, and a high uniformity of particle size. When the quantity of lime in the sinter is sufficient to replace completely separately charged limestone in blast furnaces, the sinter is called self-fluxing. Most sintering facilities still are located in blast furnace plants, despite the fact that the major portion of the raw materials used in the sintering process now comes from other locations.

Changes in the quality of sinter will be treated by documenting the changes in the proportions of materials used to produce sinter. For example, throughout this study, the relative quantity of flux used in sinter production will be used as an indicator of sinter quality. Ceteris paribus, the higher the flux content, the higher the sinter quality.

Pellets

Pelletizing was first attempted on a commercial scale in the 1920s as a means of utilizing the vast reserves of low grade (for example, an iron content of 25 percent) taconite ores in the Mesabi range. This early pelletizing process was similar to the previously described sintering process. In the first stage of production, the taconite was concentrated by crushing, washing, and magnetic separation. In the second stage, the powdered taconite concentrate was combined with coke and sintered into an agglomerate. Thus, strictly speaking, the first pelletizing facilities did not produce pellets. This operation was a commercial failure because of high costs and poor product quality.

---

*The lime, CaO, is derived from the limestone, $CaCO_3$, added to the sinter mix. In the presence of the intense heat of the sintering grate, the limestone decomposes (calcines) into lime and carbon dioxide.

After World War II, the impending depletion of domestic reserves of ore of acceptable quality (a natural iron content of at least 50 percent) led to renewed interest in the possibility of producing pellets from taconite. By 1955, the first commercially successful pellet plant was in operation. It's success resulted from the development of a mechanical process by E. W. Davis that formed the taconite concentrate into uniformly sized balls prior to hardening them in shaft furnaces or grate kilns. The physical quality of the pellets was improved substantially with the addition of the balling process.

Today, pelletizing is not restricted to the use of taconite as the basic raw material, although the other ores used are also low in quality. Because of the low quality of the ores processed, the concentration facilities should be (and are) located at mine sites to minimize the costs of shipping raw materials and disposing of enormous quantities of waste (called tailings). The agglomeration facilities are also located at mine sites. While there are a number of economies that can be achieved by locating them there, recent serious proposals to build pipelines to transport a slurry of taconite concentrate from mine sites to blast furnace plants suggest these economies are minor.* The quality of pellets has changed less over time than the quality of ores and sinter. Two major sources of pellets for domestic consumption have been the Lake Superior mines (since 1955) and Canada (since 1963). Because Canadian pellets, on the average, tend to be moderately superior in quality to U.S. pellets, the proportion of Canadian pellets consumed does affect the overall quality of this material—but most of the superiority of the Canadian product is reflected in its higher average iron content.

Time

To conveniently handle major transitions in the overall quality of the iron-bearing materials, the 1900-70 period will be divided into

---

*These economies derive from the following considerations:

(1) Land costs: they are cheaper at mine sites than in blast furnace plants.
(2) Scale economies: it is cheaper to house both concentration and agglomeration facilities in the same building than to construct two separate structures. Also, the minimum efficient scale of agglomeration operations may exceed the scale of many blast furnace plants.
(3) Inventory costs: pellets can be shipped in freezing weather because they do not freeze together, but concentrate does freeze.
(4) Quality control: the moisture content of the concentrate—crucial in the balling stage of production—may be easier to control in a continuous process in which the concentrate is agglomerated immediately.

three subperiods for analytical purposes: 1900-10, when blast furnace operators were adjusting to changing ore supplies; 1911-55, character-ized by relative uniformity in the quality of the iron-bearing inputs; and 1955-70, marked by major increases in the proportion of ores benefici-ated, major improvements in sintering and pelletizing, and large in-creases in the proportion of foreign ores consumed by the industry.

GAPS IN THE DATA ON KEY CHARACTERISTICS
OF THE IRON-BEARING MATERIALS

Even for the relatively simple quality framework of specifying the natural iron content and the proportions of scrap, ores, sinter, and pellets, the data are not complete. Generally, only the natural iron content of iron-bearing materials shipped from U.S. and Canadian mines is reported accurately by industry or government agencies. [3] These shipped materials include most ores and ore concentrates, in-cluding sinter fines, other ores consumed in sinter plants, and virtually all pellets. Excluded are scrap materials and almost all sinter (which is produced largely by the ultimate consumer, usually in facilities ad-jacent to their blast furnaces). The iron content of samples of other foreign ores is published by the U.S. Bureau of Mines in annual editions of Minerals Yearbook. On the other hand, published data on the natural iron content of domestic ores not mined in the Lake Superior region are sporadic because these mines are small enterprises generally owned by the ore user. Fortunately, from 1900 to 1955 Lake Superior mines pro-duced from 60 to 70 percent of all iron-bearing materials (including scrap) consumed in blast furnaces and supplied a large fraction of the iron materials consumed in sinter plants. The only missing data on the proportion of the iron-bearing materials consumed are that of the propor-tion of sinter consumed prior to World War II. But since sinter did not exceed ore in quality and its proportion was never above 10 percent (and well below 10 percent in most years) during this period, this is not a serious gap.

It would be possible to estimate accurately the average iron con-tent of all iron-bearing materials by dividing the total production of pig iron by the total net consumption of the iron-bearing materials if the quantity of the iron-bearing materials wasted during smelting and casting operations and the average iron content of all pig iron products were known.* From 1900 to 1970, the average iron content of all pig iron products was in a narrow range from 92 to 95 percent, with 94 per-cent representing the typical iron content of the two most important pig

---

*The formula is Natural Fe percent of Iron Materials

$$= \frac{\text{Pig Iron Produced X Average Fe Percent of Pig Iron}}{\text{Iron Material Consumed X (100 – Percent Material Wasted)}}$$

iron products, basic and Bessemer iron. The waste of iron-bearing materials changed dramatically during this period, however; and the quantity wasted per ton of pig iron produced is not known at the industry level. Hence, it will be necessary to estimate the quantity wasted to derive the average iron content of all iron-bearing materials from the input of iron-bearing materials per ton of pig iron output.

The level of waste is affected by the amount of fines in the ores, the furnace design and operating technology, and casting practices and facilities. The greater the amount of fines in the ores, the greater is the quantity of flue dust blown out of the furnace. The volume of dust, however, can be reduced if the blast furnace stack is designed to smelt ores with excessive fines (by reducing the taper between the bosh diameter and the hearth diameter) and if the top is sealed and pressurized to reduce the speed at which the air blast passes through the solid materials (thereby reducing its ability to blow small particles of ore out of the furnace). These changes in furnace design and technology were particularly important in reducing the volume of flue dust during the first two decades of this century.

Improvements in operating technology that affect the volume of flue dust include discovering the optimal blast rates for ores with varying amounts of fines, and charging the solid materials into the furnace in a certain order according to size (called "sizing"), to avoid mixing large particles with small. These improvements were made throughout the 1900-70 period, although the major impact of improved blowing practice was felt in the 1900-30 period; and the practice of sizing has been used widely only during the last 20 years.

Two major changes in casting technology that lowered the amount of pig iron wasted were the development (and diffusion) of mechanical casting (as opposed to casting iron by hand in sand molds) and the development of giant ladle cars to transport molten pig iron intended for steelmaking to steel furnaces. The latter development was an accompaniment to the dawning of "the age of steel," when steel products were displacing iron products rapidly, thus obviating the need for casting iron into pigs. These improvements in casting had a major impact on lowering waste during the first two decades of this century.

Table 5.1 summarizes the statistics that will be covered and the approximate order in which they will be discussed in this chapter. A review of the summary measures of the key characteristics of the iron-bearing materials will set the stage for more detailed investigations in later sections. Developments in the 1900-55 period and the 1955-70 period will be considered separately.

AN OVERVIEW OF CHANGES IN SUPPLY, CONSUMPTION, AND QUALITY

Table 5.2 reports the annual quantities of iron-bearing materials consumed, pig iron production, and the ratio of materials consumed to

TABLE 5.1

The Key Characteristics of the Iron-Bearing Materials

	Time Period Covered	Quantity	Iron Content Natural	Iron Content Dry	Each Material as a Proportion of
Summary Measures					
Total iron materials consumed	1900-70	X	—	—	—
Pig iron produced	1900-70	X	—	—	—
The burden rate[a]	1900-70	X	—	—	—
Sources of materials					
Lake Superior mines	1900-70	—	X	—	Total domestic materials[b]
Mesabi Range	1900-70	—	—	—	Total domestic materials[b]
Imported materials	1946-70	—	—	—	U.S. shipments, net of exports
Beneficiated domestic shipments	1946-70	—	—	—	Total domestic shipments[b]
Materials consumed in sinter plants					
All ores	1955-70	X	—	X	Total ingredients consumed[c]
Domestic	1955-70	—	—	X	Total ores consumed[c]
Foreign	1955-70	—	—	X	Total ores consumed[c]
Flue dust	1955-70	X	—	—	Total ingredients consumed[c]
Role scale	1955-70	X	—	—	Total ingredients consumed[c]
Limestone	1955-70	X	—	—	Total ingredients consumed[c]
Materials consumed in blast furnaces					
All ores	1900-70	X[e]	X[f]	—	Total materials consumed[d]
Domestic	1955-70	—	X	—	Total ores consumed[d]
Foreign	1955-70	—	X[g]	—	Total ores consumed[d]
Scrap	1900-70	X[e]	—	—	Total materials consumed[d]
Sinter	1955-70	—	—	—	Total materials consumed[d]
Pellets	1955-70	X	X	—	Total materials consumed[d]
Domestic	1955-70	—	X	—	Total pellets consumed[d]
Canadian	1963-70	—	X	—	Total pellets consumed[d]

[a] Total iron materials consumed, divided by total pig iron produced.
[b] Excluding scrap.
[c] In the sinter plant.
Note: X indicates the characteristic studied.

Source: compiled by author.

[d] In the blast furnace.
[e] Per ton of pig iron, 1900-55.
[f] Estimated in the period 1900-55.
[g] Estimated.

58

TABLE 5.2

Annual Consumption of Iron-Bearing Materials
and Production of Pig Iron: 1899-1970
(millions of gross tons)

Year	Consumption (1)	Production (2)	Burden Rate[a] (3)	Year	Consumption (1)	Production (2)	Burden Rate[a] (3)
1899	25.5	13.6	1.87	1939	59.0	31.1	1.90
1904	31.7	16.5	1.92	1940	77.8	41.1	1.89
1909	51.2	26.2	1.98	1941	93.2	49.2	1.90
1910	54.5	27.3	2.00	1942	100.8	52.8	1.91
1911	47.7	23.6	2.02	1943	103.3	54.3	1.90
1912	60.0	29.7	2.02	1944	103.0	54.5	1.89
1913	61.3	31.0	1.98	1945	90.2	47.5	1.90
1914	46.4	23.3	1.99	1946	75.7	40.0	1.89
1915	59.6	29.9	1.99	1947	99.2	52.1	1.91
1916	77.5	39.4	1.97	1948	103.4	53.6	1.93
1917	76.0	38.5	1.97	1949	91.5	47.7	1.92
1918	77.0	38.8	1.98	1950	109.7	57.7	1.90
1919	60.1	30.9	1.95	1951	118.0	62.7	1.88
1920	71.4	36.8	1.94	1952	102.6	54.7	1.88
1921	32.0	16.6	1.93	1953	125.0	66.9	1.87
1922	52.0	27.1	1.92	1954	96.2	51.8	1.86
1923	77.7	40.2	1.93	1955	127.7	68.6	1.86
1924	60.3	31.3	1.93	1956	120.7	67.0	1.80
1925	70.5	36.5	1.93	1957	126.4	70.0	1.81
1926[b]	74.6	38.7	1.93	1958	89.7	51.0	1.76
1927	69.7	35.9	1.94	1959	93.1	53.7	1.73
1928	72.1	37.4	1.93	1960	101.4	59.4	1.71
1929	80.5	41.8	1.93	1961	98.8	57.7	1.71
1930	60.0	31.0	1.93	1962	99.3	58.6	1.70
1931	33.4	18.0	1.86	1963	106.9	64.1	1.67
1932	15.0	8.6	1.75	1964	127.2	76.4	1.67
1933	24.4	13.0	1.88	1965	129.3	78.7	1.64
1934	29.5	15.7	1.88	1966	134.0	81.7	1.64
1935	39.5	20.8	1.90	1967	129.0	77.7	1.66
1936	57.1	30.2	1.89	1968	134.8	79.3	1.70
1937	68.5	36.1	1.90	1969	141.3	84.8	1.67
1938	35.4	18.5	1.91	1970	136.2	81.6	1.67

[a]Calculated by dividing column (1) by column (2).

[b]Starting in 1926, the American Iron and Steel Institute excluded iron-bearing materials used in the production of ferroalloys from the total consumption figure. Hence, production after 1926 excludes ferroalloys. Ferroalloys generally amounted to less than 1 percent of total production.

Sources: 1909-70, American Iron and Steel Institute, Annual Statistical Report (New York: AISI, 1951-70, various issues); 1899 and 1904, Census of Manufactures (Washington, D.C.: Bureau of the Census, 1899 and 1904).

iron produced (called the burden rate) in the 1900-70 period. The annual consumption of iron materials is plotted in Figure 5.1. Note the rapid rise in the consumption of these materials from 1900 to 1916 (from 25 to 78 million tons). Very large ore deposits in the Mesabi range and other Lake Superior mines made it possible to treble domestic shipments of ore in this relatively brief period of time. Except for 1921 and 1922, annual consumption of iron-bearing materials in the 1916-30 period ranged from 60 to 80 million tons; and in 10 of these 15 years, it was above 70 million tons. Following the depression decade, 1930-40, the annual consumption of iron-bearing materials reached a new high of 104 million tons in 1943 and climbed to new peaks of 118 in 1951, 125 in 1953, 128 in 1955, 134 in 1966, and 142 in 1969. From 1941 to 1970, the annual consumption dipped below 90 million tons only once, in 1946. It was this sustained, high rate of consumption that led to the impending depletion of domestic ores of acceptable natural quality, increasing dependence on both high grade foreign ores and highly beneficiated domestic ores, reported in Table 5.3.

Table 5.3 and Figure 5.2 indicate that major increases in the beneficiation of domestic ores took place in the post-World War II period, while foreign ores and agglomerates increased their share of the U.S. market from 4 percent in 1946 to a high of 38 percent in 1959. The postwar growth in beneficiation of domestic materials had three phases (see Figure 5.2). From 1946 to 1954, the proportion beneficiated rose gradually from 22 percent to 36 percent.* This increase was primarily an increase in ore concentration. From 1955 to 1963, the proportion beneficiated rose from 34 to 78 percent, reflecting a rapid growth in pellet production as well as further gains in the production and shipment of ore concentrates. After 1963, the growth in beneficiation was more gradual, reaching 92 percent in 1970.

Growth in foreign ores' share of the domestic market also occurred in three phases (see Figure 5.2). From 1946 to 1953, foreign ores (expressed as a proportion of total shipments of iron-bearing materials to blast furnace and sinter plants) rose gradually from 4 percent to 9 percent of the domestic market. From 1953 to 1959 their share rose rapidly to 38 percent of the market, but has stabilized at approximately 35 percent since 1962.

Table 5.4 and Figure 5.3—which report the major sources of domestic materials and their natural iron content—provide another focus on the pressures leading to increased beneficiation of domestic ores and importation of foreign ores during the last two decades. From 1902 to 1910, the natural iron content of Lake Superior ores fell from 55.4 to 51.7 percent, while the proportion of total domestic ores accounted for by mines in this region rose from 74 to 82 percent. The high iron content of these ores was reduced rapidly as the high grade ore deposits in the

---

*From 1929 to 1946 it rose gradually from 12 to 22 percent.

FIGURE 5.1
Annual Consumption of
Iron-Bearing Materials

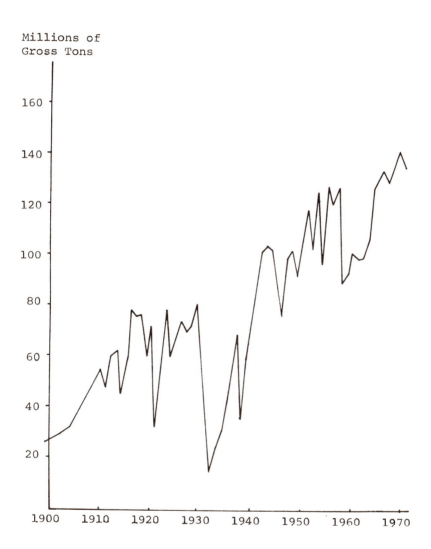

Source: Table 5.2

TABLE 5.3

Measures of Improvements in Materials Shipped
to Blast Furnace and Sinter Plants

Year	Beneficiated Domestic Materials Shipped, as a Proportion of Total U.S. Shipments[a]	Imported Materials, as a Proportion of Total Net U.S. Shipment and Imports[b]
1946	0.22	0.04
1947	0.23	0.05
1948	0.24	0.06
1949	0.24	0.08
1950	0.27	0.08
1951	0.27	0.08
1952	0.28	0.09
1953	0.30	0.09
1954	0.36	0.18
1955	0.34	0.19
1956	0.39	0.25
1957	0.40	0.25
1958	0.48	0.30
1959	0.51	0.38
1960	0.55	0.31
1961	0.64	0.28
1962	0.67	0.34
1963	0.78	0.33
1964	0.76	0.35
1965	0.77	0.37
1966	0.78	0.36
1967	0.80	0.37
1968	0.89	0.37
1969	0.89	0.33
1970	0.92	0.36

[a]These shipments include natural ores, ore concentrates, sinter, and pellets produced at mine sites but exclude scrap and sinter produced in plants distant from domestic mines.

[b]Net shipments exclude exports (which are relatively minor)

Source: Bureau of Mines, Minerals Yearbook (Washington, D.C.: U.S. Department of the Interior, various issues).

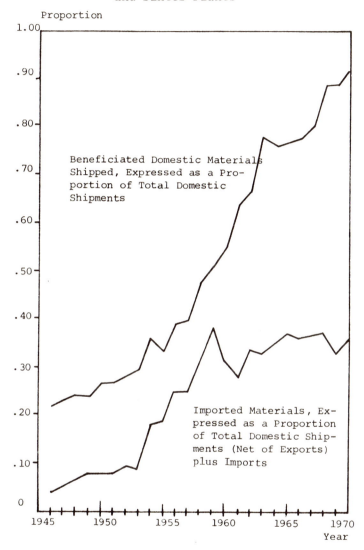

FIGURE 5.2
Improvements in Iron-Bearing Materials
Shipped to Domestic Blast Furnace
and Sinter Plants

Beneficiated Domestic Materials
Shipped, Expressed as a Pro-
portion of Total Domestic
Shipments

Imported Materials, Ex-
pressed as a Proportion
of Total Domestic Ship-
ments (Net of Exports)
plus Imports

Source: Table 5.3

TABLE 5.4

Major Sources of Domestic Iron-Bearing Materials

Year	Proportion of Domestic Shipments From		Natural Iron Percent of Lake Superior Materials	Year	Proportion of Domestic Shipments From		Natural Iron Percent of Lake Superior Materials
	Mesabi	Lake Superior			Mesabi	Lake Superior	
1900	0.30	0.74	—	1917	0.55	0.84	51.4
1901	0.32	0.74	—	1918	0.56	0.86	51.3
1902	0.36	0.76	55.4	1919	0.57	0.86	51.6
1903	0.38	0.76	54.8	1920	0.53	0.86	51.7
1904	0.42	0.73	55.0	1921	0.61	0.86	52.1
1905	0.47	0.79	54.1	1922	0.55	0.86	51.9
1906	0.50	0.80	53.4	1923	0.60	0.85	51.8
1907	0.53	0.80	52.9	1924	0.56	0.83	51.7
1908	0.49	0.78	52.6	1925	0.56	0.85	51.7
1909	0.54	0.82	51.9	1926	0.55	0.85	51.8
1910	0.53	0.82	51.7	1927	0.54	0.83	51.2
1911	0.56	0.80	51.5	1928	0.56	0.85	51.2
1912	0.57	0.85	51.7	1929	0.57	0.86	51.2
1913	0.57	0.84	51.4	1930	0.56	0.84	51.3
1914	0.53	0.82	51.3	1931	0.54	0.82	51.5
1915	0.54	0.85	51.5	1932	0.34	0.67	52.2
1916	0.54	0.85	51.2	1933	0.55	0.87	51.9

Year	Proportion of Domestic Shipments From Mesabi	Lake Superior	Natural Iron Percent of Lake Superior Materials	Year	Proportion of Domestic Shipments From Mesabi	Lake Superior	Natural Iron Percent of Lake Superior Materials
1934	0.57	0.85	51.5	1953	0.65	0.81	50.4
1935	0.56	0.84	51.4	1954	0.59	0.78	50.9
1936	n.a.	0.86	51.5	1955	0.61	0.80	50.6
1937	0.63	0.86	51.5	1956	0.60	0.78	51.3
1938	0.50	0.73	51.9	1957	0.61	0.79	52.1
1939	0.56	0.82	51.8	1958	0.60	0.77	53.8
1940	0.60	0.84	52.1	1959	0.57	0.74	53.8
1941	0.64	0.86	51.8	1960	0.62	0.78	53.8
1942	0.66	0.86	51.7	1961	0.57	0.75	55.2
1943	0.65	0.85	51.6	1962	0.60	0.76	55.6
1944	0.66	0.85	51.7	1963	0.59	0.76	56.5
1945	0.66	0.85	51.7	1964	0.56	0.75	56.8
1946	0.67	0.84	51.3	1965	0.58	0.76	56.9
1947	0.64	0.83	50.9	1966	0.58	0.77	56.8
1948	0.64	0.82	50.5	1967	0.58	0.77	57.8
1949	0.63	0.81	50.4	1968	0.60	0.77	58.7
1950	0.61	0.80	50.4	1969	0.63	0.78	59.0
1951	0.63	0.81	50.4	1970	0.63	0.78	59.3
1952	0.61	0.79	50.5				

Note: The proportions exclude scrap and include only the ore input of sinter plants located away from mines.

Source: U.S. Bureau of Mines, Minerals Yearbook (Washington, D.C.: U.S. Department of the Interior, various issues).

FIGURE 5.3

Proportions of Total Shipments of U.S. Produced
Iron Materials Accounted for by Shipments from
the Masabi Range and All Lake Superior Mines and
the Natural Iron Content of Lake Superior Shipments

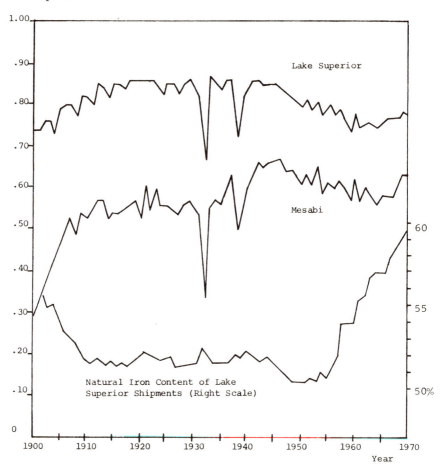

Source: Table 5.4

66

Mesabi range were exploited and production expanded. This reduced
the proportion of high grade ore from other mines in this region whose
production rates were more difficult to expand. From 1900 to 1910, the
market share of the Mesabi range rose from 30 to 53 percent in response
to a demand for iron-bearing materials that rose from 25 million to 55
million gross tons; that is, Mesabi production rose from 8 million to
30 million gross tons. From 1910 to the end of World War II, the production of Lake Superior ores and their natural iron content remained
remarkably stable; the latter was never less than 51.2 percent or
greater than 52.2 percent. This stability apparently led blast furnace
managers to expect ore shipments of this "normal" quality. But the
high consumption of ore during World War II and the postwar years
made it impossible to supply all blast furnaces with ore of this quality,
and the natural iron content of Lake Superior shipments fell from 51.7
percent in 1945 to 50.4 percent in 1949-53. The search for solutions to
the problem of declining ore quality led to an exploration for and subsequent development of foreign ore mines, an expansion of sinter plant
facilities at blast furnace sites,* and the commercialization and diffusion of pelleting facilities in the Lake Superior region. The effect of
the rapid expansion of pellet production on the natural iron content of
Lake Superior shipments was sizable; it increased from 50.6 percent in
1955 to 59.3 percent in 1970.

The effect of these changes in the Lake Superior mines also can be
seen from the perspective of the burden rate, although somewhat
opaquely (see Figure 5.4). That rate rose from 1.87 in 1899, to 2.00
in 1910, and 2.02 in 1911 and 1912. Subsequently, it fell to as low as
1.89 in the 1936-46 period. This fall does not correspond to the almost
constant natural iron content of Lake Superior ores in this period and,
thus, was probably a result of reductions in the waste of these materials in the blast furnace process. Finally, from 1948 to 1966, the burden
rate fell from 1.93 to 1.64, rising slightly in the 1967-70 period.

THE 1900-55 PERIOD

The data on sources and quality reported in Table 5.4 and the
burden rate reported in Table 5.2 provide the major sources of information concerning the quality of the iron-bearing materials in this period.
Accordingly, it will be assumed that the average iron content of all
ores (including sinter) was constant (at 51.5 percent) in the 1911-55
period; and efforts will be concentrated on determining the reduction

---

*Sinter plants consumed 15 percent of natural ores and ore concen-
trates from domestic and foreign sources in the period 1951-56. This
proportion rose to 20 percent in 1957, 28 percent in 1958, 37 percent
in 1959, and stabilized at almost 50 percent in the period 1960-70.

FIGURE 5.4

The Burden Rate

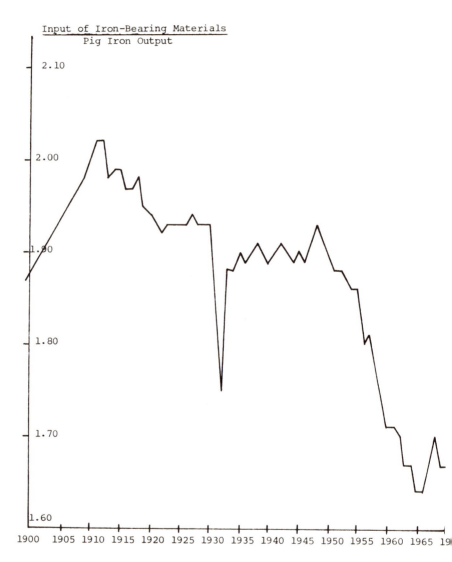

Input of Iron-Bearing Materials
Pig Iron Output

Source: Table 5.2

in waste of the iron-bearing materials in the 1900-55 period and the iron content of scrap in the same period.*

The Early Period: 1900-11

Scrap consumed per ton of pig iron produced (the scrap rate) remained relatively constant at 0.09 to 0.11 in 1899, 1904, and 1909, rising to 0.15 in 1911. Ore consumed per ton of pig iron produced (the ore rate) rose from 1.96 in 1899 to 1.81 in 1904, 1.89 in 1909, and declined to 1.87 in 1911.[4] Hence, the burden rate rose from 1.87 in 1899 to 2.02 in 1911, an increase of 8 percent. Assuming that the decline in the natural iron content of Lake Superior ores, from 56 percent in 1899 to 51.5 percent in 1911, was representative of all domestic ores consumed in this period† and assuming further that the iron content of scrap was constant and reasonably close to the iron content of ores in this period, then the proportion of iron materials wasted in this period was constant (because the decline in the iron content of ores consumed completely accounted for the rise in the burden rate). Since Mesabi ores (which have large amounts of fines) accounted for less than 30 percent of ores consumed in 1899 and 56 percent of ores in 1911, these statistics indicate that blast furnace operators were learning to reduce ore losses (in the form of flue dust) in this period. The problem was solved partly by designing most of the new furnaces constructed in this period—more than 100—to accommodate the Mesabi ores.

The 1911-55 Period

The reduction in waste and the average iron content of scrap during this period will be estimated by regression analysis. Because of the strongly atypical nature of the 1931-34 subperiod when rates of capacity

---

*See earlier sections of this chapter for a review of technological factors affecting the proportion of iron-bearing materials wasted.

†Since relatively fixed proportions of eastern ores (New York, Pennsylvania, and New Jersey) and southern ores (Alabama) provided the 20 to 25 percent of ores not supplied by Lake Superior mines and because these ores were relatively constant in quality, the decline in the natural iron content was probably a bit smaller.

utilization were extremely low, it was deleted from the main body of the analysis. During that period, blast furnaces were charged with high quality ores and unusually large amounts of scrap; and they were operated at low driving rates. All of these factors undoubtedly lowered the volume of flue dust and greatly lowered the burden rate (see Figures 5.4 and 5.5).

The subperiod 1948-53 also could be deleted from the main body of the analysis on the grounds that natural ores and ore concentrates shipped from domestic mines were unusually low in quality. But since sinter production was expanded significantly from 1951-53, offsetting the lower iron content of the ores, only the 1948-50 subperiod was deleted. (Since no distinction between sinter and ores is being made during this period, increased sinter production had the effect of increasing the average iron content of iron materials, excluding scrap.)

As a deductive conjecture, the evidence from the literature suggests that the iron content of scrap should be in the 45 to 75 percent range, with an average of 60 percent being likely. A simple regression of the ore rate against the scrap rate gave the following results:

$$\hat{OR} = 2.091 - 2.031 \; (SR), \quad R^2 = 0.717 \tag{5.1}$$

where: OR = the ore rate
SR = the scrap rate

The slope parameter value (-2.031) is impossibly large; it suggests that either the iron content of ore was well below 51.5 percent or the iron content of scrap was well above 75 percent.

This result was obtained because the specification of equation 5.1 is incomplete due to the omission of variables capable of explaining improvements in the efficiency of ore utilization. The simplest method by which these improvements can be estimated is to expand the specification of equation 5.1 to include dummy variables representing subperiods in the total 1911-55 period.[*] Dummy variables, in effect, periodically will change the value of the intercept parameter and—if properly selected—will allow the slope parameter to attain its proper value. The fluctuations in the burden rate suggest that the 1911-55 period should be separated into five subperiods (see Figure 5.5), 1911-12; 1913-18; 1919-30; 1935-47; and 1951-55.

Including dummy variables in the regression equation for these five subperiods gave the following results:

$$\hat{OR} = 2.028 - 0.039X_1 - 0.085X_2 - 0.117X_3 - 0.147X_4$$
$$- 1.069 \; (SR), \tag{5.2}$$

---

[*]An alternative technique would be to incorporate time as an explanatory variable. This method requires, however, that the rate of improvement in efficiency of ore utilization be constant, which is unduly restrictive.

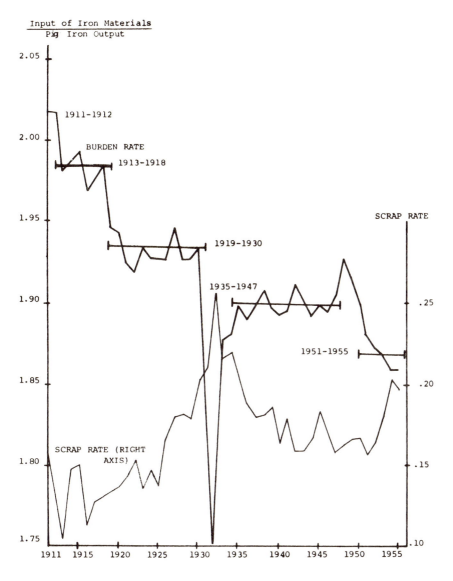

FIGURE 5.5

The Burden Rate and Scrap Rate

Source: American Iron and Steel Institute, Annual Statistical Report, various issues.

71

$$R^2 = 0.985$$

where: $X_1 = 1$, 1913-18; $= 0$, otherwise

$X_2 = 1$, 1919-30; $= 0$, otherwise

$X_3 = 1$, 1935-47; $= 0$, otherwise

$X_4 = 1$, 1951-55; $= 0$, otherwise

The 99 percent confidence interval for the slope parameter value is
$(-0.871, -1.267)$, which is equivalent to an iron content range for
scrap of (44.9 percent, 65.3 percent). The parameter value itself
indicates the average iron content of scrap was 55.1 percent if that of
ore were 51.5 percent. All the dummy variable parameters are significantly different from each other at the 99 percent confidence level.

Equation 5.2 gives a reasonable tradeoff between scrap and ore.
All parameter values are highly significant, yielding strong evidence of
falling dust rates and declining iron losses in the casting operation
over the whole period. These improvements are equivalent to saving
294 pounds of ore (± 19 pounds at the 99 percent level of confidence)
per ton of pig iron. Sixty percent of this improvement occurred prior to
1930, as would be expected.

Estimation of the ore-scrap tradeoff for the 1951-55 period alone
gave the following results:

$$\hat{OR} = 1.953 - 1.464\,(SR), \quad R^2 = 0.999 \tag{5.3}$$

Equation 5.3 indicates a jump in the iron content of scrap. Since the
iron content of ore and sinter averaged about 51 percent, the average
iron content of scrap was 74 percent. The 99 percent confidence
interval for the slope parameter is $(-1.443, -1.485)$, which translates
to an iron content range of (72.9 percent, 75.1 percent) for scrap. This
range does not overlap the 99 percent confidence interval for the entire
period 1911-55, implying a significant increase in the iron content of
scrap. Evidence from a major producer over the 1953-70 period supports
this finding.[5]

If the 1935-47 period is analyzed separately, the following is the
estimated equation.

$$\hat{OR} = 1.926 - 1.154\,(SR), \quad R^2 = 0.860 \tag{5.4}$$

The average iron content of scrap in this period was 59.5 percent. The
99 percent confidence interval for this mean value is (39.4 percent,
79.4 percent), indicating the average iron content of scrap during
1935-47 was not significantly different from the average scrap quality
for the whole period.

In summary, the iron content of scrap averaged 55 percent for the
entire 1911-55 period. Since this figure can be placed at 74 percent

for the last four years of this period (1951-55), the implication is that it averaged 53 percent for the 1911-47 period. Increases in the physical efficiency of the smelting and casting operations reduced the burden rate by 0.15 tons of standard quality ore from 1911 to 1955.

THE RECENT YEARS: 1955-70

Materials Consumed in Sinter Plants

Table 5.5 reports the relative importance of domestic versus foreign mines in supplying sinter plants[6] (not located at mine sites)* with sinter fines and other ores and the average dry iron content of the ores from each source. Domestic ore supplied more than 80 percent of the ores consumed in 1955, but this proportion fell rapidly to 60 percent in 1957-61 and more gradually to 40 percent in 1968-70.

The dry iron content of total ores consumed rose by 6.2 percent for two reasons.† First, the iron content of ores from domestic mines rose by 4.3 percent and the dry iron content of foreign ores rose by 2.7 percent in this period. Second, the iron content of foreign ores ranged from 4 to 7 percent higher than that of domestic ores.

The proportion of key ingredients consumed in sinter plants annually from 1957 to 1970, reported in Table 5.6 and illustrated in Figure 5.6, indicates that ore was the major input. Its proportion of total input rose from 69 percent in 1957 to 78 percent in 1960, remained relatively constant from 1961 to 1965 at 76 percent, then gradually declined to 69 percent in 1970. The other sources of iron were scale and flue dust. The proportion of scale remained relatively constant at 2 percent until 1962, gradually rising thereafter to $7\frac{1}{2}$ percent in 1970. The proportion of flue dust exhibited a major decline from 27 percent in 1957 to 13 percent in 1959, then fell more gradually to $6\frac{1}{2}$ percent in 1970. The practice of adding flux to other sinter ingredients to produce self-fluxing sinter grew rapidly, as the proportion of flux rose from 1 percent in 1957 to 17 percent in 1970.

The shifts in the proportions of these ingredients contributed to the quality of sinter. The increase in the proportion of scale—which is essentially oxidized iron scrap (and therefore high in iron content)—and the decrease in the proportion of flue dust—which detracts from the physical properties of sinter—were definite improvements. The increase in the proportion of flux may have improved the physical quality of sinter and reduced the thermal load of blast furnaces, because

---

*Sinter produced at mine sites rarely exceeded 3 million gross tons during this period.

†Since the dry iron content of some ores consumed in sinter plants had to be estimated, the calculated dry iron content of total ores consumed is subject to error. The maximum possible error is ±2 percent in any given year.

TABLE 5.5

Estimated Dry Iron Content of Ore
Consumed in U.S. Sinter Plants

Year	Domestic Ores Proportion	Dry Iron (in percent)	Foreign Ores Proportion	Dry Iron (in percent)	Total Ores Dry Iron (in percent)
1955	0.84	55.7	0.16	62.4	56.8
1956	0.70	55.2	0.30	61.8	57.2
1957	0.59	55.1	0.41	62.1	58.0
1958	0.55	56.9	0.45	62.1	59.2
1959	0.61	56.7	0.39	62.1	58.8
1960	0.62	56.6	0.38	62.1	58.7
1961	0.60	56.2	0.40	61.4	58.3
1962	0.55	57.6	0.45	62.9	60.0
1963	0.50	58.3	0.50	63.5	60.9
1964	0.46	58.5	0.54	63.6	61.3
1965	0.44	59.3	0.56	63.4	61.6
1966	0.45	59.1	0.55	64.0	61.8
1967	0.43	58.7	0.57	64.4	61.9
1968	0.40	59.3	0.60	64.9	62.7
1969	0.40	59.4	0.60	65.2	62.8
1970	0.40	60.0	0.60	65.1	63.0

Note: Figures exclude sinter produced at mine sites.

Sources: American Iron Ore Association, Iron Ore (Cleveland: AIOA, various issues). The dry iron content of foreign ores was estimated from data reported in Iron Ore, data from Gerald Manner, The Changing World Market for Iron Ore 1950-1980 (Baltimore: The Johns Hopkins Press for Resources for the Future, 1969), and data from U.S. Bureau of Mines Minerals Yearbook (Washington, D.C.: U.S. Department of the Interior, various issues).

TABLE 5.6

Materials Consumed in Sinter Plants: Quantities and Proportions
(in millions of gross tons)

Year	Ore Q	Ore P	Flue Dust Q	Flue Dust P	Scale Q	Scale P	Flux Q	Flux P	Q Total[a]	Sinter Production[a]
1955	16.2	—	n.a.	—	n.a.	—	n.a.	—	n.a.	20
1956	17.0	—	n.a.	—	n.a.	—	n.a.	—	n.a.	20
1957	20.2	0.694	7.9	0.271	0.6	0.021	0.04	0.014	29.9	24.4
1958	19.3	0.715	5.7	0.211	0.6	0.022	1.4	0.052	27.0	22.8
1959	27.1	0.770	4.7	0.134	0.9	0.026	2.5	0.071	35.2	27.7
1960	36.4	0.783	5.3	0.114	1.0	0.022	3.8	0.082	46.5	37.4
1961	34.6	0.750	5.1	0.111	1.1	0.024	5.3	0.115	46.1	37.7
1962	34.5	0.758	4.6	0.101	1.3	0.029	5.1	0.112	45.5	38.9
1963	37.8	0.761	4.5	0.091	1.7	0.034	5.7	0.115	49.7	41.0
1964	40.5	0.761	4.6	0.086	2.2	0.041	5.9	0.111	53.2	44.9
1965	40.0	0.755	4.4	0.083	2.6	0.049	6.0	0.113	53.0	44.1
1966	39.2	0.734	4.2	0.079	2.9	0.054	7.1	0.133	53.4	45.0
1967	36.1	0.722	3.7	0.074	3.1	0.062	7.1	0.142	50.0	42.8
1968	33.7	0.709	3.1	0.065	3.1	0.065	7.6	0.160	47.5	41.0
1969	33.9	0.705	3.0	0.062	3.3	0.069	7.9	0.164	48.1	41.3
1970	31.0	0.694	2.9	0.065	3.3	0.074	7.5	0.168	44.7	38.4

[a]Located away from mines. Sinter production at mine sites varied from 2.2 to 3.8 million tons during this period.

Note: Q = quantity; P = proportion; n.a. = not available.

Sources: Ore and sinter production, American Iron Ore Association, Iron Ore (Cleveland: AIOA, various issues). Flue dust, scale, and flux, American Iron and Steel Institute, Annual Statistical Report (New York: AISI, various issues).

FIGURE 5.6

Proportions of Materials Consumed in
Sinter Plants

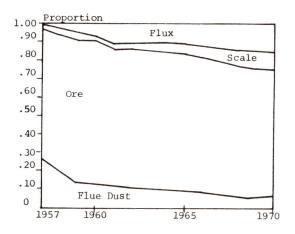

FIGURE 5.7

Materials Consumed and Sinter Produced

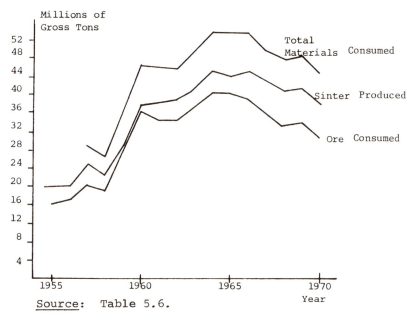

this flux was calcined on the sintering grate, a process that reduces its weight by approximately one-half and eliminates the need to calcine an equal weight of flux in blast furnaces.

Because ingredients other than ore are used in sinter production, the annual volume produced is greater than the ore input, as illustrated in Figure 5.7. The gap between sinter production and ore consumption narrowed to less than one million tons in 1959 and 1960—the years in which the proportion of total inputs accounted for by ore was at its peak—but subsequently widened to over seven million tons in 1968, 1969, and 1970, as the volume of scale and flux rose by six million tons.

Because some of the sinter ingredients are reduced in weight during the sintering process as gases are driven off, the quantity of inputs was greater annually than the volume of sinter produced. Flue dust is reduced in weight to the extent that it contains coke particles and moisture. The flux is reduced in weight as it is calcined, and ore is reduced by an amount equivalent to its water content. By assuming that flux suffers a 50 percent weight loss and flue dust a 25 percent weight loss during the sintering process, the difference between materials throughput and sinter output can be narrowed to a number that approximates the water content of the ore.* By dividing this number by the weight of the ore input, the average water content of the ore input can be estimated. The resulting figures are 15 percent for 1957-60, 11 percent for 1961-65, and 7 percent for 1966-70. The difference between these three figures seems large enough to warrant the claim that the ore inputs to U.S. sintering plants became drier over time, reducing the work load on the sintering grate. These periodic reductions in water content are attributable to changes in the quality of domestic and Canadian ores. Two trends can be associated with shipments of domestic and Canadian ores over time: an increase in the relative volume of sinter fines and an increase in the portion that was concentrated.

Materials Consumed in Blast Furnaces

Sinter

Although it was possible to show that the dry iron content of ores consumed in sinter plants increased significantly, it is not possible to determine accurately the iron content of the finished product because of ignorance about the quality of other materials consumed.

---

*The fraction of coke charged into blast furnaces that is blown out is smaller than the counterpart fraction of ore, but the relative position of these fractions may have changed during this period in response to changes in the composition of materials charged into blast furnaces. Hence, this number only crudely approximates the water content of ore.

Ores

It is evident from Table 5.7 that foreign ores were not substituted for domestic ores as rapidly or extensively as they were in sinter plants. The total consumption of foreign ores in blast furances rose from 16 percent in 1955 to a high of 36 percent in 1969. The natural iron content of domestic and foreign ores rose by 3 and 4.4 percent, respectively; and the iron content of foreign ores ranged from $4\frac{1}{2}$ to 7 percent above domestic ores. Accordingly, the natural iron content of total ores consumed in U.S. blast furnaces rose by 4.5 percent in this period.

Since an increasing proportion of domestic and Canadian ores were concentrated prior to shipment, it is possible that the water content of the total ores consumed fell during this period—as discovered in the case of ores consumed in sinter plants.* If correct, this implies that the rise in the natural iron content overstates the fall in the proportion of dry impurities in these ores and, in this sense, the improvement in the quality of ores consumed.

Pellets

Table 5.8 reports the natural iron content of pellets consumed in blast furnaces from 1961 to 1970. Prior to 1961, the iron content of pellets was about 61.5 percent.† From 1961 to 1970, there was a steady increase in the iron content of pellets consumed, from 61.4 percent to 63 percent. This increase was caused by two factors. First, beginning in 1963 the domestic industry began to import pellets from Canada, and by 1966 Canadian pellets were supplying 18 percent of the total pellets consumed, maintaining this share in the next four years. Since Canadian pellets were roughly two percent higher in iron content than domestic pellets, this substitution had the effect of raising the natural iron content of pellets consumed by 0.4 percent. Second, the iron content of domestic pellets rose gradually by 1.2 percent during this decade.

Scrap

Direct information on the iron content of scrap is not reported by any federal agency or trade association. Hence, the iron content of scrap had to be estimated indrectly. The evidence from the preceding section indicated that the iron content of scrap was 74 percent in the 1951-55 period. A major integrated steel firm ("company X") supplied

---

*The U.S. and Canadian proportion (combined) fell from 90 percent in 1955 to 70 percent in 1970. The water content of other foreign ore was below that of Canadian and domestic supplies.

†Based on data supplied by a major integrated steel firm that requested anonymity, and is therefore referred to as "company X" in this study.

TABLE 5.7

Estimated Natural Iron Content of Ore
Consumed in U.S. Blast Furnaces

	Domestic Ores		Foreign Ores		Total Ores
Year	Proportion	Natural Iron Percent	Proportion	Natural Iron Percent	Natural Iron Percent
1955	0.84	50.7	0.16	57.0	51.7
1956	0.81	50.3	0.19	57.2	51.6
1957	0.81	50.2	0.19	57.3	51.6
1958	0.79	52.0	0.21	57.7	53.2
1959	0.77	51.9	0.23	57.4	53.2
1960	0.69	51.9	0.31	57.7	53.7
1961	0.76	52.1	0.24	57.7	53.5
1962	0.73	53.2	0.27	58.1	54.5
1963	0.74	53.3	0.26	58.6	54.7
1964	0.71	53.4	0.29	58.4	54.9
1965	0.69	53.9	0.31	58.3	55.3
1966	0.68	53.6	0.32	59.1	55.5
1967	0.67	53.3	0.33	59.2	55.4
1968	0.72	53.6	0.28	59.0	55.2
1969	0.64	53.6	0.36	60.7	56.1
1970	0.67	53.7	0.33	61.4	56.2

Source: American Iron Ore Association, Iron Ore (Cleveland: AIOA, various issues). The natural iron content of foreign ores was estimated from data reported in Iron Ore; Gerald Manner, The Changing World Market for Iron Ore 1950-1980 (Baltimore: The Johns Hopkins Press for Resources for the Future, 1969); and U.S. Bureau of Mines, Minerals Yearbook (Washington, D.C.: U.S. Department of the Interior, various issues).

TABLE 5.8

Estimated Natural Iron Content of Pellets
Consumed in U.S. Blast Furnaces

Year	U.S. Pellets Proportion	Natural Iron Percent	Canadian Pellets Proportion	Natural Iron Percent	Natural Iron Percent of Pellets Consumed
1961[*]	1.000	61.4	0	—	61.4
1962	1.000	61.5	0	—	61.5
1963	0.961	61.7	0.039	65.5	61.9
1964	0.891	62.0	0.109	64.7	62.3
1965	0.853	62.1	0.147	64.7	62.4
1966	0.819	61.9	0.181	64.5	62.4
1967	0.825	62.3	0.175	64.3	62.6
1968	0.817	62.5	0.183	64.2	62.8
1969	0.860	62.7	0.140	64.4	62.9
1970	0.821	62.6	0.179	64.2	63.0

[*]The natural iron content of domestic pellets from 1955 to 1960
was approximately 61.5 percent.

Source: American Iron Ore Association, Iron Ore (Cleveland: AIOA,
various issues).

data that also indicate (indirectly) that the iron content of their scrap
averaged 74 percent during 1953-70.

Changes in Their Proportions

The quantities and proportions of scrap, sinter, pellets, and ores
consumed in U. S. blast furnaces are reported in Table 5.9. The proportion of ores declined from 71 percent in 1955 to 24 percent in 1970
(see Figure 5.8), as ores were supplanted by sinter and pellets. From
1955 to 1960, the sinter proportion grew more rapidly than the pellet
proportion, rising from 17 percent to 42 percent, while the latter rose
from 1.5 percent to 10 percent. During 1960-63 the sinter proportion
remained at 42 percent, while the pellet proportion expanded to
17 percent. After 1963, pellets displaced both ores and sinter in
relative importance; their proportion reached 40 percent by 1970,
while the sinter proportion fell to 30 percent.

As illustrated in Figure 5.9, the major changes in consumption
of these materials occurred in distinct patterns. During the 1958 re-
cession, the consumption of iron-bearing materials fell by 25 percent,
equivalent to almost 30 million tons. Natural ores and ore concentrates

TABLE 5.9

Materials Consumed in U.S. Blast Furnaces: Quantities and Proportions
(in millions of gross tons)

Year	Scrap		Sinter		Pellets		Ores		Total
	Q	P	Q	P	Q	P	Q	P	
1955	13.6	0.106	22.1[a]	0.173	1.9	0.015	90.0	0.706	127.7
1956	11.5	0.095	21.4	0.178	3.5	0.029	84.0	0.698	120.7
1957	11.5	0.091	30.2	0.236	4.9	0.039	80.2	0.634	126.4
1958	7.9	0.089	26.9	0.292	7.2	0.080	48.5	0.539	89.7
1959	7.8	0.084	32.0	0.341	7.6	0.081	45.8	0.494	93.1
1960	8.0	0.080	42.1	0.422	9.6	0.097	40.5	0.401	101.4
1961	7.8	0.078	41.8	0.418	13.5	0.135	36.4	0.369	98.8
1962	8.0	0.081	43.0	0.441	13.7	0.140	33.5	0.338	99.3
1963	8.8	0.082	45.3	0.419	18.0	0.167	35.5	0.332	106.9
1964	10.2	0.080	49.6	0.390	26.7	0.210	40.7	0.320	127.2
1965	10.1	0.077	48.3	0.372	30.7	0.237	40.5	0.314	129.3
1966	8.5	0.064	49.6	0.369	37.1	0.276	39.0	0.291	134.0
1967	7.9	0.061	48.0	0.372	39.1	0.303	34.0	0.264	129.0
1968	7.4	0.055	45.7	0.339	44.8	0.332	36.9	0.274	134.8
1969	8.6	0.059	44.5	0.315	52.2	0.370	36.2	0.256	141.3
1970	7.8	0.057	41.0	0.301	54.9	0.403	32.6	0.239	136.2

[a]The quantity of sinter consumed is as much as 10 million tons higher than sinter production at plants not located at mine sites because small amounts of sinter are produced at mines and shipped to blast furnaces, small quantitites are imported from Canada, and inventories changed.

Note: Q = quantity; P = proportion.

Sources: American Iron and Steel Institute, Annual Statistical Report (New York: AISI, various issues); and American Iron Ore Association, Iron Ore (Cleveland: AIOA, various issues).

FIGURE 5.8

Proportion of Materials Consumed in
Blast Furnaces

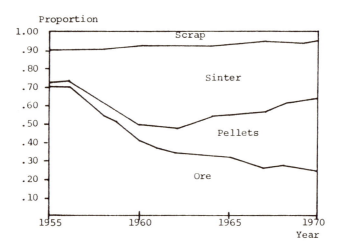

FIGURE 5.9

Quantities of Materials Consumed in
Blast Furnaces

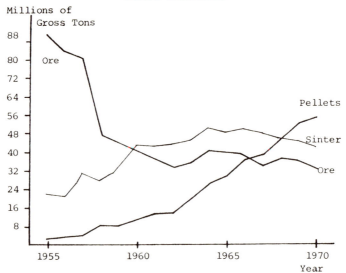

Source: Table 5.9.

82

absorbed the entire drop and continued to decline (at a much less rapid pace) during the "low demand" years 1959-62, as they were replaced by agglomerates. Sinter production almost doubled from 1955 to 1960, rose gradually from 42 to 50 million tons in the mid-1960s, and fell gradually back to 41 million tons in 1970. Pellet consumption increased gradually from 1955 to 1962 and more rapidly thereafter, reaching 55 million tons in 1970. Ore consumption fluctuated similarly to sinter consumption after 1962, but remained about 9 million tons below sinter consumption.

Estimated Average Natural Iron Content of All Materials Consumed

The major barrier to be surmounted in estimating the average natural iron content of the total iron-bearing materials is the uncertainty surrounding the natural iron content of sinter. After attempting a number of more sophisticated techniques, it was decided to estimate the iron content of all materials by simply multiplying the inverse of the burden rate by 100 percent. The results are reported in Table 5.10 for the 1950-70 period. If the average iron content of pig iron is 94 percent, the accuracy of this estimating process requires that the average efficiency at which these materials are utilized in blast furnace plants is also 94 percent.

TABLE 5.10

Estimates of the Iron Content of All Materials
Consumed in Blast Furnaces

Year	Estimated Natural Iron Content	Year	Estimated Natural Iron Content
1950	52.6	1961	58.5
1951	53.2	1962	59.0
1952	53.3	1963	60.0
1953	53.5	1964	60.1
1954	53.8	1965	60.9
1955	53.8	1966	60.9
1956	55.5	1967	60.2
1957	55.3	1968	58.8
1958	56.9	1969	60.1
1959	57.7	1970	60.0
1960	58.5		

Note: Derived from column 3 of Table 5.2 by multiplying the inverse of the burden rate by 100.

Although it seems unlikely that blast furnace consumption efficiency was uniformly 94 percent in this period, there is conflicting evidence on this point. Table 5.6 indicates that flue dust consumption in sinter plants fell from 8 million tons in 1957 to 3 million tons in 1968-70. If the annual consumption of this material is accurately correlated with its production, then consumption efficiency increased over time, which is appealing. On the other hand, evidence supplied by company X indicated that there was no trend in consumption efficiency over time, despite a major substitution of agglomerates for ore. Consumption efficiency fluctuated between 87 percent and 98 percent, averaging 93+ percent from 1953 to 1970 in company X's blast furnaces.

A tentative check on the accuracy of the estimated natural iron content of all iron-bearing materials consumed can be made by employing (1) the proportions for pellets, sinter, scrap, and ore (reported in Table 5.9); (2) the iron content for ore and pellets consumed (reported in Tables 5.7 and 5.8, respectively); and (3) the estimated iron content of all materials consumed (reported in Table 5.10) to estimate an iron content for sinter. If the iron content of scrap remained uniformly at 75 percent, then the estimated average iron content of all iron materials consumed implies that the iron content of sinter was 60.0 percent in 1956, 58.8 percent in 1962, and 56.0 percent in 1970. Adjusting these figures by removing the calcined flux content of sinter indicates that the iron content of unfluxed sinter would have been 60.4 percent, 62.5 percent, and 61.7 percent, respectively, in these three years. Although the indicated drop from 1962 to 1970 contradicts the earlier evidence that the dry iron content of ore consumed in sinter plants rose from 60 percent to 63 percent, it is not sufficiently contradictory to warrant scrapping the estimated natural iron content of all iron materials consumed because the estimated iron content of all such blast furnace materials are subject to a number of errors.

NOTES

1. The technical characteristics of the iron-bearing materials have been summarized from various issues of "The Making, Shaping and Treating of Steel" (Pittsburgh: United States Steel Company, published at irregular intervals from 1920 to 1966).

2. Webster's New World Dictionary (Springfield, Mass.: G. & C. Merriam Co., 1968).

3. The U. S. Bureau of Mines, the American Iron Ore Association, and the American Iron and Steel Institute are three important sources of data.

4. Data for 1899, 1904, and 1909 were reported in the U. S. Census of Manufactures (Washington, D. C.: U. S. Department of Commerce, Bureau of the Census) for those years. The American Iron

and Steel Institute's Annual Statistical Report (New York: AISI) provided the figures for 1911.

5. See Chapter 6 for a discussion of the data provided by company X.

6. These data are reported in Myles G. Boylan, "The Economics of Changes in the Scale of Production in the U. S. Iron and Steel Industry from 1900 to 1970" (Case Western Reserve University, Ph. D. dissertation, 1973). The data used to calculate the proportions and dry iron contents of foreign and domestic ores consumed in sinter plants and blast furnaces were much less aggregative than the figures reported in Tables 5. 5 and 5. 7.

6

**THE SOURCES OF
ENERGY AND THE
FLUXING AGENTS**

BACKGROUND INFORMATION

Energy and Flux and Their Roles
in the Blast Furnace Process

The fluxing agents, limestone, and dolomite (but chiefly limestone, $CaCO_3$) are related functionally to the iron-bearing materials and the energy inputs as a complementary input. They are needed to separate the impurities in the iron materials and coke from the molten iron in the hearth of the furnace. The energy inputs are chiefly coke and recycled blast furnace gas (used to heat the hot blast stoves); but small amounts of charcoal and raw coal were used in the early part of the century, replaced by injected fuels (natural gas, fuel oil, coal tar, and coal dust) during the 1960s.* These inputs are classified in this study as energy inputs because they are the source of heat needed inside the furnace to melt the fluxing agents and iron-bearing materials. But they also fill the role of reducing agents, needed to combine with the oxygen contained in the iron oxides; and for this purpose they could be classified as materials.

---

*Energy is also consumed to operate skip hoists, cranes, ladle cars, and other equipment ancillary to blast furnace operations. These energy uses will not be considered because of the lack of data and the assurances from industry experts that energy consumption for these purposes is relatively small.

The energy inputs can be substituted for each other, but only to a limited extent. For example, increased hot blast temperature can be substituted for fossil fuels until either the hot blast stoves, pipes, and lower parts of the blast furnace can no longer withstand further increases in temperature or until further declines in fossil fuel inputs cannot be sustained because it is necessary for a certain minimum amount of hydrocarbons to be present to reduce the iron oxides. Similarly, injected fuels can be substituted for coke. But injecting fuels into the air blast lowers the flame temperature in the hearth of the furnace unless compensating increases in the hot blast temperature are employed. Thus, since major reductions in the flame temperature cannot be allowed without affecting adversely the blast furnace process, this substitution eventually will strain the ability of the stoves, pipes, and such, to withstand further increases in the hot blast temperature.

The Basic Plan of Analysis

The major portion of this chapter is designed to explain changes in consumption of coke per ton of iron produced. Those furnaces that used primarily charcoal or raw coal in the early part of the century have been deleted from this study. The analysis of changes in the level of coke consumption will be undertaken in two sections: the 1910-55 period, utilizing results from the last chapter; and the 1950-70 period,[*] using more detail because of the substantial changes in the quality of the iron-bearing materials documented in the previous chapter.

The analysis of changes in the consumption of fluxing agents per ton of iron produced is included with the analysis of coke in this chapter, primarily for technical reasons rather than intrinsic economic interest. That is, although the cost of flux is only a minor component of the total materials cost, flux and coke are strong complementary inputs; and the determinants of the level of coke consumption and flux consumption are almost identical.

Changes in the Levels of Energy and Flux Consumption

Table 6.1 and Figure 6.1 indicate the changes from 1911 to 1970 in the consumption of coke and flux per ton of iron produced. (Hereafter,

---

[*]The overlap with the earlier period is on purpose and is designed to give two views of the transitory nature of the period of overlap, 1950-55.

TABLE 6.1

The Coke Rate and Flux Rate in the 1911-70 Period

Year	Coke Rate[a]	Flux Rate[b]	Year	Coke Rate[a]	Flux Rate[b]	
1911	1.075	0.511	1941	0.873	0.362	
1912	1.075	0.508	1942	0.896	0.387	
1913	1.073	0.519	1943	0.900	0.387	
1914	1.038	0.496	1944	0.905	0.398	
1915	0.992	0.494	1945	0.920	0.392	
1916	1.007	0.465	1946	0.934	0.406	
1917	1.032	0.482	1947	0.950	0.412	
1918	1.047	0.495	1948	0.954	0.440	
1919	1.018	0.487	1949	0.935	0.428	
1920	1.016	0.485	1950	0.922	0.429	
1921	0.970	0.442	1951	0.924	0.432	
1922	0.958	0.433	1952	0.922	0.426	
1923	0.988	0.447	1953	0.906	0.418	
1924	0.975	0.420	1954	0.873	0.395	
1925	0.947	0.415	1955	0.873	0.387	0.003
1926	0.920	0.397	1956	0.850	0.362	0.004
1927	0.935	0.393	1957	0.842	0.350	0.005
1928	0.919	0.382	1958	0.799	0.301	0.028
1929	0.907	0.374	1959	0.785	0.267	0.047
1930	0.900	0.358	1960	0.749	0.239	0.064
1931	0.884	0.346	1961	0.708	0.210	0.091
1932	0.872	0.345	1962	0.690	0.195	0.087
1933	0.864	0.360[c]	1963	0.669	0.195	0.089
1934	0.888	0.360[c]	1964	0.655	0.196	0.078
1935	0.871	0.362	1965	0.656	0.203	0.076
1936	0.895	0.362	1966	0.641	0.181	0.087
1937	0.903	0.361	1967	0.631	0.166	0.091
1938	0.887	0.347	1968	0.624	0.146	0.095
1939	0.880	0.355	1969	0.626	0.151	0.093
1940	0.878	0.362	1970	0.630	0.153	0.093

[a]For pig iron production only. From 1911 to 1935, the American Iron and Steel Institute reported coke rates for pig iron and ferroalloys jointly. From 1926 to 1932, coke rates were reported both for pig iron separately and pig iron plus ferroalloys. The latter, broader category had coke rates from 0.012 to 0.015 higher than pig iron alone. Hence, 0.013 was subtracted from the reported coke rates in the 1911-25 figures.

[b]Prior to 1955, all flux was charged directly into blast furnaces as a separate material. Starting in 1955, a portion of the flux input was combined with other ingredients in sinter plants and "indirectly charged" into blast furnaces as part of the sinter input.

[c]Estimated.

FIGURE 6.1

Unit Input Rates of Coke and Flux

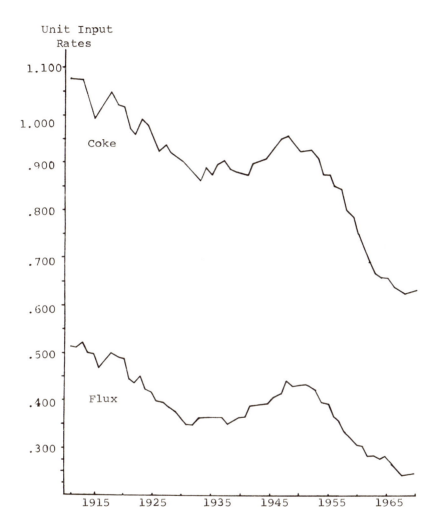

Note: Tons of input per ton of iron. Flux includes both directly and indirectly charged fluxing materials.
Source: Table 6.1.

these input-output ratios will be referred to as the coke rate and flux rate, respectively.) The coke rate fell by 20 percent from 1911 to 1933, from 1.075 to 0.862, rising annually only in 1915, 1917, 1918, 1923, and 1927. The flux rate fell by 30 percent during the same period, from 0.508 to 0.360, rising only in 1913, 1917, 1918, 1923 and 1933. The rises in both of these input rates during World War I and 1923 can be explained by the high utilization of capacity during those years, necessitating the use of nearly obsolete furnaces with poor top design and low blast temperatures. The 1933-41 period was characterized by little trend in both series. Except for 1938, capacity utilization rose gradually throughout this period from very low levels. From 1941 to 1948, the coke rate rose steadily by 9 percent, from 0.873 to 0.953; and the flux rate rose more rapidly by 22 percent, from 0.362 to 0.440. This period was characterized by high utilization rates and a modest deterioration in ore quality. Finally, from 1948 to 1968 the coke rate declined steadily by 35 percent, from 0.953 to 0.625; and the flux rate fell by 45 percent, from 0.440 to 0.240. Major improvements in the chemical and physical quality of the iron-bearing materials took place during these two decades.[*]

Determinants of the Levels of Coke and Flux Consumption: Technical Considerations

The determinants of the flux rate[1] are the levels of impurities in the iron-bearing materials and coke. The natural iron content of the iron-bearing materials is a partial determinant of the level of impurities in the iron-bearing materials, but not a complete one due to variations in the water content of these materials and variations in the composition

---

[*]Statistically the close correlation between coke and flux is supported by the following regression equations.

$$\frac{1910\text{-}55}{CR} = 0.48 + 1.13 \text{ (FR)} - 0.04X, \quad R^2 = 0.95$$

$$\frac{1950\text{-}70}{CR} = 0.47 + 1.07 \text{ (FR)}, \quad R^2 = 0.98$$

FR is the flux rate, CR is the coke rate, and X equals one in 1951-55 and zero in other years. The relationship between coke and flux did not change noticeably between 1910-55 and 1950-70. These equations indicate that the relative fluctuations in the flux rate have been greater than the relative fluctuations in the coke rate.

of the impurities.* Accordingly, it is necessary to specify the composition of the iron-bearing materials (that is, scrap, pellets, sinter, and ore) to adjust for variations in the water content and composition of impurities. The level of impurities in coke varies between 12 percent and 8 percent, with 10 percent representing the median proportion. (The chief impurity in coke is ash. Traces of sulfur are also present.)

The determinants of the coke rate, given the temperature of the air blast and the rate of consumption of other fuels, are the qualities and quantities of the iron-bearing materials, the flux per ton of iron produced, and, possibly, the average size of all utilized furnaces in the industry. Each of the materials requires a certain irreducible quantity of coke to supply sufficient heat to melt it. Additionally, the iron oxides require a sufficient volume of carbon to reduce them to pure iron. The coke rate may fall for increases in furnace size because the area of the shell of the furnace increases more slowly than the working volume of the furnace, potentially reducing heat losses.

To understand the role of coke, consider the possible reactions of coke with other substances as it descends into the hotter part of the blast furnace stack.

(1) It can react with carbon dioxide, forming carbon monoxide but absorbing heat (almost five times as much as is yielded by burning coke per unit of carbon).
(2) It can react with iron oxide (direct reduction), yielding almost pure iron and carbon monoxide but absorbing heat (two and one-half times the heat yielded by burning coke per unit of carbon).
(3) It can react with oxygen (burn), yielding carbon monoxide and heat.

Part of the carbon monoxide derived from the above three basic reactions reacts with iron oxides (indirect reduction), yielding almost pure iron, carbon dioxide, and a modest amount of heat (one-fifth the amount derived by burning coke per unit of carbon). It would be ideal for energy conservation if all of the carbon monoxide needed to reduce the iron oxides were supplied by the combustion of coke and if coke reacted only with oxygen.

For a number of reasons, this ideal state of affairs is not possible to achieve. Some coke will reduce iron oxides directly or react with carbon dioxide, thereby increasing the coke rate above its theoretically lowest level.

---

*The impurities in the iron-bearing materials that require fluxing are (in descending order of magnitude) silica, alumina, and sulfur. Silica is clearly an acid, and alumina has some properties of acids. The flux (a base) combines readily with these impurities and forms a molten layer of slag over the pool of molten iron. The sulfur combines less readily with the fluxing agents, although it can be removed by maintaining a sufficiently thick layer of slag. Generally, the raw flux requirement is specified to be 2X (silica + alumina).

In addition to indirect reduction and the air blast, carbon dioxide is supplied by the calcination of flux (which in its natural state is approximately one-half carbon dioxide by weight). Thus, there are at least three basic reasons why the coke rate and the flux rate are correlated strongly. Heat is needed to melt the flux. The carbon dioxide given off during calcination is as likely to react with coke as the carbon dioxide resulting from indirect reduction (thereby increasing the coke rate). Finally, the flux rate is a good indicator of the chemical (and a partial indicator of the physical) quality of the iron-bearing materials and, hence, can be expected to predict the level of coke needed to reduce and melt these materials with reasonable accuracy.

To establish a direction of causality in the relationships between coke, flux, and the iron-bearing materials for the statistical analyses in later sections of this chapter, the following considerations are helpful:

(1) The flux rate is determined primarily by the burden rate (the total input of iron-bearing materials per ton of iron) and the proportions of the various iron-bearing materials consumed. To the extent that impurities in the coke determine the flux rate, it is assumed that variations in the coke rate are fundamentally due to variations in the burden rate.

(2) The coke rate is determined jointly by the burden rate, the proportions of the various iron-bearing materials consumed, and the flux rate. Thus, since the flux rate is determined by the burden rate, the coke rate may be treated as fundamentally dependent on the burden rate (including the composition of the burden).

COKE AND FLUX RATES FROM 1910 TO 1955:
A STATISTICAL ANALYSIS

The Relationship Between the Iron-Bearing Materials, Flux, and Coke

In the previous chapter, it was discovered that ore and scrap were close substitutes in terms of iron content. It would be incorrect, however, to conclude on the basis of this finding that they have similar effects on the coke rate and the flux rate. In periods during which the scrap rate increased (and the ore rate declined), the coke and flux rates should have been reduced because scrap is partly comprised of open-hearth slag, which contains high proportions of magnesia and lime—the primary fluxing agents—and because scrap contains a much smaller proportion of the kinds of impurities that require fluxing.

Statistical analysis confirms this expectation. If scrap and ore had almost identical effects on the coke and flux rates, then the burden rate should explain statistically at least as much of the variance in the coke rate and flux rate as the ore rate alone. The following two regressions* where $X_4 = 1$, 1951-55 and $X_4 = 0$ otherwise,

$$\hat{CR} = -1.984 + 1.512(BR) + 0.043X_4, \quad R^2 = 0.804 \tag{6.1a}$$

$$\hat{FR} = -2.063 + 1.285(BR) + 0.074X_4, \quad R^2 = 0.754 \tag{6.2a}$$

compared with the counterpart equations where the ore rate is substituted for the burden rate,

$$\hat{CR} = -0.956 + 1.076(OR) + 0.038X_4, \quad R^2 = 0.906 \tag{6.1b}$$

$$\hat{FR} = -1.222 + 0.926(OR) + 0.071X_4, \quad R^2 = 0.890 \tag{6.2b}$$

suggest that ore and scrap are not near substitutes. The R^2 is substantially higher in the latter two equations, by 10 percent and 14 percent, respectively.

To integrate this analysis with the last chapter, the predicted value of the ore rate, $\hat{OR}$, will be used as the explanatory variable (see equation 5.2). By regressing the coke rate and the flux rate against the predicted value of this variable, instead of the true ore rate (which was done in equations 6.1b and 6.2b), the R^2 is not lowered by much.

$$\hat{CR} = -0.964 + 1.081(\hat{OR}) + 0.038X_4, \quad R^2 = 0.897 \tag{6.1c}$$
$$\text{(versus } 0.906)$$

$$\hat{FR} = -1.221 + 0.925(\hat{OR}) + 0.071X_4, \quad R^2 = 0.871 \tag{6.2c}$$
$$\text{(versus } 0.890)$$

R^2 is lowered by 0.009 and 0.019, respectively.[†] Note that the parameter values are affected little by the substitution of $\hat{OR}$ for OR. By substituting the equation for $\hat{OR}$ into these two equations, the following two equations are obtained:

---

*The values of these variables during the subperiods 1931-34 and 1948-50 have been deleted from the regression equations in this section in accordance with the format established in the previous chapter.

[†]Of the variance in the coke rate, 10.3 percent is not explained by fluctuations in OR or X_4. Sixty-four percent of this unexplained variance is explained by the deviation of the actual flux rate from the flux rate value predicted by equation 6.2c. That is,

$$\frac{\Lambda}{CR - \hat{CR}} = 0.856 \; (FR - \hat{FR}), \quad R^2 = 0.64.$$

Thus, it may be argued that the coke rate is not determined perfectly by OR and X_4 because the flux rate is not determined perfectly by these same variables.

$$\hat{CR} = (1.228 - 0.042X_1 - 0.092X_2 - 0.127X_3 - 0.121X_4) - 1.156(SR)$$
$$(6.1d)$$

$$\hat{FR} = (0.665 - 0.036X_1 - 0.079X_2 - 0.108X_3 - 0.065X_4) - 0.989(SR)$$
$$(6.2d)$$

In the equations above:

$X_1 = 1$, 1913-18; $= 0$ otherwise

$X_2 = 1$, 1919-30; $= 0$ otherwise

$X_3 = 1$, 1935-47; $= 0$ otherwise

$X_4 = 1$, 1951-55; $= 0$ otherwise

$SR =$ the scrap rate

The format of these equations is in accordance with the substitution equation 5.2 developed in the last chapter.

These equations may be interpreted as an integrated system. The explanatory power of the independent variables in equation 6.1d depends in part on their explanatory power in equation 6.2d. For example, increases in the scrap rate caused larger absolute decreases in the coke rate, and the other way around, partly because such increases lowered the required level of flux and partly because the dust rate was thereby lowered.* In addition, the periodic reductions in the coke rate indicated by the parameters of the four dummy variables in equation 6.2d may be attributed, in part, to each of the following:
(1) reductions in the dust rate noted in the previous chapter;
(2) increases in the hot blast temperature; and
(3) periodic reductions in the flux rate, which in turn may have been due to increased operating skill by blast furnace managers.†

---

*The flue dust blown out of blast furnaces is comprised of both small ore particles and coke particles. Flux and scrap generally are charged in large enough pieces to avoid being blown out.

†An alternative view of the interdependent relationship of equation 6.1d to equation 6.2d can be derived from the relationship between the residual (error) terms of these two regressions:

$$\frac{\wedge}{CR - \hat{CR}} = 0.856\,(FR - \hat{FR}), \quad R_2 = 0.64$$

This equation may be rewritten.

$$\overset{*}{CR} = [\hat{CR} - 0.856(\hat{FR})] + 0.856(FR)$$

Substituting equation 6.1c for $\hat{CR}$ and equation 6.2c for $\hat{FR}$ yields the following equation:

$$\hat{CR} = \overset{*}{CR} = 0.081 + 0.289(\hat{OR}) - 0.023X_4 + 0.856(FR)$$

The periodic reductions in the coke rate and flux rate (given a constant value of the scrap rate) were reversed partly in the 1951-55 period, indicating that period's transitional nature. Despite a decline in the average burden rate of 0.030 between 1935-47 and 1951-55, the coke rate and flux rate increased by 0.006 and 0.043, respectively. The primary causes of these divergent movements were changes in the quality of iron-bearing materials. First, the average moisture content of the burden materials was declining due to increases in the proportions of sinter and beneficiated ore. To the extent the lowering of the burden rate is due to this factor, it would have little effect on the flux rate. Second, the very high iron content of sinter (18.3 percent of the burden) and scrap (9.6 percent of the burden) tends to cloud the low quality of the ore (72 percent of the burden). It is possible that the level of impurities per ton of iron rose, despite a decrease of 0.030 in the burden rate in this period. Last, there is the strong possibility that some blast furnace operators were overfluxing their furnaces to improve furnace productivity by increasing the voidage (the empty spaces) in the blast furnace stack and, thus, the ability of the air blast to penetrate the materials in the stack. None of these explanations can be sufficient alone, but taken together they constitute a plausible explanation of the atypical nature of the 1951-55 period.

The Dominant Role of Coke

A question was raised earlier in this chapter concerning which of the two functions served by coke in the blast furnace process (heat or reduction) is dominant. To explore this point in greater detail, consider the following two equations:

$$\hat{CR} = 0.003 + 0.963(FR) + 0.282(BR) - 0.028X_4, \quad R_2 = 0.960 \quad (6.3a)$$

$$\hat{CR} = -0.687 + 0.697(BR + FR) + 0.000X_4, \quad R^2 = 0.947 \quad (6.3b)$$

Equation 6.3a supports the energy dominance thesis. The intercept term is almost zero, implying that, to reduce the coke rate to zero, the

---

This is equivalent to the regression of CR on $\hat{OR}$, X_4 and FR with $R_2 = 0.96$. Finally, substituting for $\hat{OR}$ yields the following equations:

$$\hat{CR} = 0.667 - 0.011X_1 - 0.025X_2 - 0.034X_3 - 0.065X_4 - 0.309(SR)$$
$$+ 0.856(FR)$$

The parameters of the four dummy variables in this equation should represent periodic reductions in the coke rate attributable to reductions in the dust rate and increases in the hot blast temperature.

flux rate and the burden rate would have to be zero as well. The fact that the flux rate parameter is much larger than the burden rate parameter is in keeping with an earlier contention that the flux rate is a better predictor of the coke rate than is the burden rate. It also supports the fact that flux is more difficult to melt than ore. The parameter value of the dummy variable, X_4, for the 1951-54 period indicates a decrease in the coke rate not explained by concomitant changes in the flux rate or burden rate. This may be explained by the same factors previously cited about the atypical nature of the 1951-55 period.

Equation 6.3b supports the reduction dominance thesis. This equation is very close to the form:

$$CR = b(BR + FR - 1),$$

where "b" is a positive constant. This form implies the coke rate would be zero if the sum of all other solid materials per ton of pig iron were equal to one. In this circumstance, there would be no impurities requiring fluxing; and there would be no oxides of iron requiring reduction. In essence, the inputs would be pure iron. The parameters of this equation, however, can be adjusted to support the energy dominance thesis by considering the changes that occurred in the average hot blast temperature.

From 1911 to 1955, the average temperature of the air blast rose from 25 percent to 40 percent. Since the hot blast supplied approximately 15 percent of the energy input in 1911, this rise supports a drop of $4\frac{1}{2}$ to 7 percent in the coke rate. The flux rate plus burden rate fell by 11 percent, while the coke rate fell by 18 percent in this period. Hence, the fall in the coke rate, net of the substitution of the hot blast temperature for coke, could have been exactly proportionate to the fall in the consumption rate of the other solid materials—supporting the energy dominance thesis—and it certainly was not large enough to support the reduction dominance thesis.

Thus, it appears most likely that the energy role of coke dominated its reduction role. Personal discussions with industry experts found support for this conclusion. For example, William Collison, a practicing blast furnace engineer employed by the Arthur G. McKee Company, stated in a 1970 interview that furnaces were quite efficient in reducing iron oxides, but inefficient as vehicles for heat transfer.

Summary of the 1910-55 Period

The basic model of Chapter 3 requires fixed input-output ratios for materials, as long as the work done by the activity is unchanged. In this and the previous chapter, it has been shown that the model's requirement basically was fulfilled in the blast furnace industry in the 1910-55 period. The work done by the blast furnace industry would have been constant had it not been for

(1) the opportunity to substitute scrap for ore, -0.041

(2) some variance in the quality of the scrap, and

(3) the dust rate losses and the losses of iron in the casthouse that were controlled gradually over the period.

Justifications were presented for classifying coke primarily as an energy source (of heat) as opposed to a material (for reducing iron oxides). The implication of this finding is that fluctuations in the coke rate are caused by fluctuations in the unit energy requirement of the transformation process.

A STATISTICAL ANALYSIS OF THE COKE AND FLUX RATES: 1950-70

Introduction

There are two approaches that could be adopted in the analysis of the 1950-70 period. Either the agglomeration activity could be incorporated with the smelting activity, so the basic iron-bearing input continued to be crude ore; or the effects of agglomerated iron-bearing inputs on the smelting activity could be estimated. The second approach has been followed for three reasons:

(1) Incorporating the agglomeration activity with the smelting activity would not leave the work done by this larger activity sequence unchanged in comparison with the smelting activity prior to 1950. This follows from the substantial changes that occurred in the nature of the crude ore input and the large increase in the share of foreign ores.

(2) Part of the agglomeration activity, mainly pelletizing, occurs at sites physically and (at least partially) managerially separated from the smelting activity.

(3) In terms of developments in blast furnace scale and blast furnace capital requirements, the effects of agglomerates on blast furnace productivity need to be estimated regardless of which approach is adopted.

In addition to improvements in the iron-bearing materials, the coke rate and flux rate were influenced by another improvement in materials that began in 1955: the practice of adding flux materials to sinter. This practice clearly has the effect of lowering the direct flux rate (that is, the input of flux charged directly into the furnace). In addition, it lowers the coke rate because the sinter flux is calcined during the process of producing sinter, thereby reducing the weight of the sinter flux by roughly 50 percent and reducing the quantity of carbon dioxide produced in the blast furnace. According to the United States Steel Corporation, replacing one ton of directly charged flux with sinter flux saves, on the average, 400 pounds of coke. [2] For this reason the

TABLE 6.2

The Composition and Natural Iron Content of All Iron-Bearing Materials
Consumed Annually in U.S. Blast Furnaces: 1950-70

| | Percentage of the Total Iron-Bearing Materials Accounted for by | | | | | Natural Iron Percent |
Year	Scrap, Scale, Cinder, and Such	Calcined Flux in the Sinter	Sinter, Net of Calcined Flux	Pellets	Direct Ores and Ore Concentrates	of the Iron-Bearing Burden*
1950	8.7	0	15.5	0	75.8	52.6
1951	8.3	0	17.8	0	73.9	53.2
1952	8.8	0	17.5	0	73.7	53.3
1953	9.5	0	19.2	0	71.3	53.5
1954	10.9	0	21.8	0	67.3	53.8
1955	10.6	0.1	17.2	1.5	70.6	53.8
1956	9.5	0.1	17.7	2.9	69.8	55.5
1957	9.1	0.2	23.4	3.9	63.4	55.3
1958	8.9	0.8	28.4	8.0	53.9	56.9
1959	8.4	1.4	32.7	8.1	49.4	57.7
1960	8.0	1.9	40.3	9.7	40.1	58.5
1961	7.8	2.7	39.1	13.5	36.9	58.5
1962	8.1	2.6	41.5	14.0	33.8	59.0
1963	8.2	2.7	39.2	16.7	33.2	60.0
1964	8.0	2.4	36.6	21.0	32.0	60.1
1965	7.7	2.3	34.9	23.7	31.4	60.9
1966	6.4	2.7	34.2	27.6	29.1	60.9
1967	6.1	2.8	34.4	30.3	26.4	60.2
1968	5.5	2.8	31.1	33.2	27.4	58.8
1969	5.9	2.8	28.7	37.0	25.6	60.1
1970	5.7	2.8	27.3	40.3	23.9	60.0

*Calculated, assuming 94 percent blast furnace conversion efficiency and 94 percent Fe pig iron.
Source: American Iron and Steel Institute, Annual Statistical Reports (New York: AISI, 1950-70).

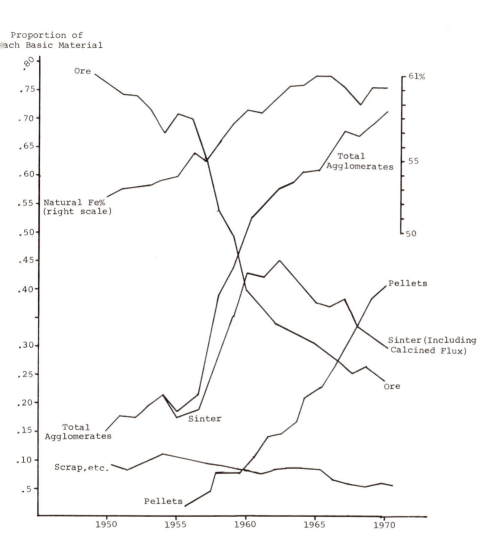

FIGURE 6.2

The Composition and Natural Iron Content of Iron-
Bearing Materials Consumed in U.S. Blast Furnaces:1950-70

Source: Table 6.2

99

indirect flux rate (that is, the sinter flux rate) will be included in the equations explaining the changes in the direct flux rate and the coke rate.

Table 6.2 and Figure 6.2 illustrate the changes in the key independent variables of this analysis. From 1950 to 1955, the natural iron content and the proportion of agglomerates and scrap rose slowly. In 1956, the natural iron content jumped by $1\frac{3}{4}$ to $55\frac{1}{2}$ percent, reflecting a major improvement in the iron content of domestic ores and a large increase in the consumption of foreign ores. Thereafter, the natural iron content and the proportion of agglomerates rose very rapidly, both reaching 61 percent in 1963. The natural iron content appears (temporarily) to have stabilized at 60 percent during the second half of the 1960s, despite further (although slower) increases in the proportion of agglomerates to 70 percent by 1970. This was caused by a drop of 2 percent in the scrap proportion and an increase in the proportion of flux in sinter.

It also should be noted that the proportion of pellets increased steadily from 1.5 percent in 1955 to 40 percent in 1970, and the sinter proportion peaked at 44 percent in 1962 and has since fallen to 30 percent in 1970.

A Statistical Analysis of Industry Level Data

The investigation of the relative impact of the various improvements in the iron-bearing materials and such on the direct flux rate and coke rate begins by utilizing industry level data. Subsequently, the results of this analysis will be compared with results based on data pertaining to individual blast furnace operations; and then the two sets of findings will be integrated.

The Flux Rate

The direct flux rate was found to be strongly dependent on the quality of the iron-bearing materials.

$$\overset{*}{DFR} = 0.018 + 0.253\ (\frac{100}{Fe\ \%}) - 0.561\ (\frac{scrap\ \%}{100}) - 0.288\ (\frac{pellet\ \%}{100})$$
$$-0.166\ (\frac{sinter\ \%}{100}) - IFR, \quad R^2 > 0.97 \qquad (6.4a)$$

where

$$\overset{*}{DFR} = \hat{TFR} - \overset{*}{IFR}$$

---

*This approach was used instead of directly regressing "DFR" because the direct approach resulted in a slope parameter value for IFR of -0.732 instead of the theoretically correct value -1.

$\hat{\text{TFR}}$ = the least-squares predicted value of the total flux rate

IFR = the indirect flux rate

This equation can be altered slightly to incorporate the additional information that indirect flux is a sinter ingredient. Defining "d" as the diffusion of calcined flux in sinter,*

$$d = \frac{\text{calcined flux } \%}{\text{sinter } \%} \times 100$$

it can be shown that[†]

$$\hat{\text{DFR}} = 0.018 + 0.253 \left(\frac{100}{\text{Fe } \%}\right) - 0.561 \left(\frac{\text{scrap } \%}{100}\right) - 0.288 \left(\frac{\text{pellet } \%}{100}\right)$$

$$- (0.166 + 0.033d) \left(\frac{\text{sinter } \%}{100}\right) \tag{6.4b}$$

Equation 6.4b imparts the following information:

(1) Increases in the iron content (or reductions in the burden rate = [100/Fe %]) lower the direct flux rate. Indeed, reductions in the burden rate caused almost exactly proportionate reductions in the direct flux rate since 1950.

(2) Scrap was almost twice as effective as pellets, and pellets were almost twice as effective as unfluxed sinter in lowering the direct flux rate.[‡] But when the self-fluxing power of sinter reached "d = 3.9 percent" (its average value), sinter gained the edge over pellets.

(3) However, if an average furnace were charged with 100 percent sinter with an iron content of 55 percent and a calcined flux content of 9 percent (equal to "d" in this example), there would be no need to directly charge flux and the indirect flux rate would be approximately 0.300. This may be contrasted against the hypothetical situation in which this average furnace is charged with 100 percent pellets with an iron content of 60 percent, resulting in a direct flux rate of only 0.150. Indeed, charging this average furnace completely with an unfluxed sinter burden with an iron content of 60 percent results in a direct flux rate of "only" 0.270; that is, 10 percent less than the total flux

---

*"Calcined flux percent" and "sinter percent" refer to proportions of these inputs in the total iron-bearing burden. See Table 6.2.

[†]The key step is the almost constant relationship between the IFR and the calcined flux content of the burden: IFR = 0.033 (calcined flux percent).

[‡]Based on the direct regression equations for $\hat{\text{DFR}}$ and $\hat{\text{TFR}}$ (see equation 6.4a), scrap was not significant, though all other variables were significant at the 90 percent level.

requirement of the highly fluxed sinter. A complete ore burden with an iron content of 60 percent (that is, a high grade ore) would require a direct flux rate of 0.440. This is the benchmark against which the other results should be contrasted.

While equation 6.4b includes all the major characteristics of the iron-bearing materials—characteristics to which the actual reductions in the total and direct flux rates should be attributed—it is missing certain key variables that could explain time-related improvements in the physical quality (uniformity of size) of each basic kind of material and in operating practice (such as sizing the materials). If these improvements were made steadily throughout the 1950-70 period, then the effectiveness of pellets in reducing the flux rate would tend to be overstated (and the effectiveness of ore understated), while the calculated effectiveness of sinter is approximately correct.

The results notes in (2) and (3) above seem to confirm this educated guess. Although some reductions in the flux rate can be achieved with materials possessing superior physical properties, it seems unlikely that it would fall from 0.440 to 0.150 (with the iron content held constant at 60 percent) when pellets were substituted completely for ore.

The Coke Rate

Similar results were obtained for the coke rate (standard errors are included in brackets).[*]

$$CR = 0.402 + 0.346 \, (\frac{100}{Fe \, \%}) - 1.336 \, (\frac{scrap \, \%}{100}) - 0.450 \, (\frac{pellet \, \%}{100})$$
$$[0.127] \qquad\qquad [0.421] \qquad\qquad [0.108]$$

$$- (0.123 + 0.027d) \, (\frac{sinter \, \%}{100}), \; R^2 = 0.993$$
$$[0.143] \, [0.014] \qquad\qquad\qquad\qquad\qquad (6.5)$$

It may be noted that

(1) increasing the iron content of the burden is a highly significant method of decreasing the coke rate;
(2) increasing the proportion of scrap is the most effective way to lower the coke rate, followed in descending order by pellets, fluxed sinter, unfluxed sinter, and ore;

---

[*]If the coke rate is first regressed against the natural iron content, the proportions of the iron-bearing materials in the burden, and D$\hat{F}$R, and then the right-hand side of equation 6.4b is substituted for D$\hat{F}$R almost exactly the same results are produced as are found in equation 6.5. (Exactly the same results would have been produced if D$\hat{F}$R had been used in place of D$\hat{F}$R.)

(3) the stronger effectiveness of scrap versus pellets is significant at the 95 percent level of confidence;

(4) the stronger effectiveness of pellets over completely fluxed sinter (d = 9 percent) is not significant, but pellets are more effective in lowering the coke rate than moderately fluxed sinter (d = 4 percent) at the 85 percent level of confidence and more effective than unfluxed sinter at the 95 percent level of confidence; and

(5) unfluxed sinter is not more effective than ore in reducing the coke rate at an acceptable level of confidence. But moderately fluxed sinter is more effective than ore at the 90 percent level of confidence.

Table 6.3 illustrates the changes in the coke rate that can be expected when the natural iron content fluctuates in the range (50 percent, 65 percent). A change of 10 percent in the natural iron content of the burden in the relevant range (50 percent, 65 percent) causes a change in the opposite direction of 0.100 to 0.120 in the coke rate. Reductions in the coke rate due to increases in the natural iron content are slightly less than proportionate to the reductions in the burden rate (100/Fe %) when the increases in the natural iron content are achieved by substituting other materials for low grade ore. This finding is consistent with the fact that such substitutions lower the water content as well as the impurity content of the burden.

The estimated effect of sinter flux on the coke rate is reported in Table 6.4. Fairly dramatic reductions in the coke rate result when sinter flux is substituted for directly charged flux. In the limit, the coke rate is reduced by 22 percent (from 0.837 to 0.649) when completely self-fluxing sinter (d = 9 percent) is substituted for unfluxed sinter. This reduction is almost three times larger than the estimates made by the United States Steel Corporation.[3] In reference to the last figure in the right-hand column in Table 6.4, United States Steel estimated the reduction in the coke rate to be 0.064 (against 0.188). Since the estimated effect of changes in the iron content on the coke rate seems reasonable, the error is probably concentrated in the parameter value of "d" in equation 6.5; that is, this parameter value should be -0.013 instead of -0.027.[*]

The ranking of the various iron-bearing materials in terms of their effectiveness in reducing the coke rate seems correct. It is known, for example, that a high level of uniformity in the size of the iron-bearing particles causes a greater portion of these materials to be reduced indirectly by carbon monoxide gas by promoting an even circulation of this gas throughout the entire burden.[4] If this gas is channeled through only a portion of the burden due to size heterogeneity, then the

---

[*]If the parameter value of "d" in equation 6.5 were changed to -0.013, then the effect of substituting completely self-fluxing sinter for unfluxed sinter in the lower right hand column of Table 6.4 would be 0.117 - 0.055 = 0.062.

TABLE 6.3

Decreases in the Coke Rate Caused by Increases in the
Natural Iron Content from a Base Value of 50 Percent

Iron Percent	51.5	53	54.5	56	57.5	59	60.5	62	63.5	65
CR Decrease	-0.021	-0.41	-0.059	-0.077	-0.094	-0.110	-0.125	-0.139	-0.153	-0.166

TABLE 6.4

Decreases in the Coke Rate Caused by Increases in the
Calcined Flux Content of Sinter[a,b]

Fluxed Sinter, Expressed as a Proportion of the Total Burden[c] (in percent)	Calcined Flux Content of Fluxed Sinter		
	$d = 3$ (percent)	$d = 6$ (percent)	$d = 9$ (percent)
33	0.027 - 0.006 = 0.021	0.054 - 0.012 = 0.042	0.081 - 0.018 = 0.063
67	0.054 - 0.012 = 0.042	0.108 - 0.024 = 0.084	0.162 - 0.037 = 0.125
100	0.081 - 0.018 = 0.063	0.162 - 0.037 = 0.125	0.243 - 0.055 = 0.188

[a]From a base value of zero.
[b]Assuming unfluxed sinter has an iron content of 62 percent. The results are reported in the form:
(decrease in CR due to increase in d) minus (increase in CR resulting from decreased iron content due to increase in d). The iron content of a complete sinter burden when $d = 9$ percent is 56.4 percent.
[c]If the total burden initially had been unfluxed sinter, then the coke rate would have been 0.837 prior to the substitution of fluxed sinter.
Source: Equation 6.5.

remaining (protected) portion of the burden is likely to reach the hotter part of the furnace in its oxidized state and can be reduced directly. Also, the smaller the size of the individual particles (to a minimum diameter of about five-eighths inches), the greater is the ratio of surface area to volume, increasing the likelihood that the iron oxides will be reduced indirectly before reaching the portion of the furnace hot enough to sustain direct reduction. Pellets possess these properties more completely than sinter, and sinter more completely than ore.

Nevertheless, it is still likely that the effectiveness of pellets (and scrap) as against ores in reducing the coke rate is overstated in the same way that the effectiveness of sinter flux was overstated in equation 6.5. The reasons are not hard to find. Equation 6.5 lacks variables capable of including the effects of (1) increases in blast furnace size; (2) increases in the average temperature of the air blast; and (3) the diffusion of the practice of injecting fuels into the blast furnace through the air blast; on the coke rate. Although it can be argued that the substitution of agglomerates for ores made it more attractive economically to increase blast temperatures and substitute injected fuels for coke (because the greater uniformity of particle size of the agglomerates allowed the tempo of the reactions in the furnaces to be increased), the strong association between the increasing consumption of agglomerates and the displacement of coke by other fuels does not mean that the latter practice was necessary. Accordingly we now turn to a brief review of the probable effects of increased size, increased blast temperature, and fuel injection on the coke rate.

Blast Furnace Size and Coke Consumption: The Case of Company X

A major integrated steel producer, hereafter designated as company X to comply with the request of a senior officer that the company remain anonymous, provided operating data for all of their blast furnaces for selected years in the period 1953-70. Individual observations are on operating results for one blast furnace year, with further disaggregation if a given blast furnace produced more than one type of iron during any year. The data cover the number of days during each year in which each furnace was operated, the size of each furnace (measured by the hearth diameter of the furnace), the composition and natural iron content of the iron-bearing materials consumed by each furnace, and the coke rate of each furnace. Data on furnaces operating in the south (on southern ores) and furnaces producing ferroalloys were removed. The remaining data provided 149 observations of a pooled time series-cross sectional nature. Forty-nine of these observations cover the operations of blast furnaces producing basic pig iron on a sustained basis in vertically integrated plants.

This data pool was not wholly representative of the industry. Company X switched to higher proportions of agglomerates (particularly pellets), but achieved smaller reductions in their average coke rate than the industry as a whole during this period.* The improvements made in their stock of blast furnaces were below the industry average.

To obtain an estimate of the average iron content of the ore, scrap, sinter, and pellets consumed by company X, their data (149 observations) were employed to estimate the following equation:

$$\hat{Fe}\% = 50.8 + 23.3 \left(\frac{scrap\ \%}{100}\right) + 10.5 \left(\frac{pellet\ \%}{100}\right) + 12.9 \left(\frac{sinter\ \%}{100}\right)$$

$$R^2 = 0.836$$

The parameter estimates were each significant at the 99 percent level of confidence. The implication of this equation is that the ore employed by company X was of relatively low quality, with an iron content of only 50.8 percent. The average iron content of scrap, pellets, and sinter was 74.1 percent, 61.3 percent, and 63.7 percent, respectively. The value for sinter implies that company X lagged behind the industry in the use of self-fluxing sinter.

In designing the regression equation to effectively utilize company X's data, the form of the industry level equation was not duplicated. The indirect flux rate had to be dropped because no information on flux rates were provided. Given the high average iron content of their sinter, this could matter very little. Two other variables were included, one for the length of time the blast furnace was operated between shutdowns, if this period was less than one year, and another to indicate the size of the furnace (measured by hearth diameter). Each observation in the larger set of 149 was weighted by the tonnage produced during the length of time covered by the observation, up to one year, to achieve estimates consistent with those based on industry level data. The following results were obtained:

$$\hat{CR} = 0.537 + 0.380 \left(\frac{100}{Fe\ \%}\right) - 0.285 \left(\frac{scrap\ \%}{100}\right) - 0.230 \left(\frac{pellet\ \%}{100}\right)$$

$$- 0.163 \left(\frac{sinter\ \%}{100}\right) - 0.015\ (OP) - 0.008\ (HD),\quad R^2 = 0.824\ (6:6)$$

where OP (the period of operation) = 1: 10 days-1 month
                                    = 2: 1-3 months

---

*One possible explanation is that company X was one of the first to build a major pelletizing facility; and, as a result, the quality of the pellets was unquestionably below average. One private research foundation places the quality of most of their pellet production in the bottom 10 percent of total U.S. and Canadian pellet production, where quality is based on measures of physical strength (that is, resistance to degradation) and impurities.

$$= 3: \ 3\text{-}6 \text{ months}$$
$$= 4: \ 6\text{-}10 \text{ months}$$
$$= 5: \ 10 \text{ months-}1 \text{ year,}$$

and HD = the hearth diameter (measured in feet).

All parameter values were found to highly significant.

The fact that the period of operation is so important reflects its correlation with a number of factors that are missing from the equation. The furnaces that are operated for only short periods of time tend to be older (and smaller), generally are used only in periods of peak demand, often are used to make relatively small batches of special grades of iron, commonly have less efficient ancillary equipment (such as high pressure tops, hot blast stoves, fuel injection devices, turboblowers, and computerized control), and probably are operated by less able managers, since it would be most likely that the most capable managers are assigned to furnaces that are operated for a full year. The significance of the period of operation also can be traced to the fact that blowing-in (starting) and blowing-out (stopping) a blast furnace is costly in terms of coke consumption. Thus, the shorter the period of operation, the higher the average coke rate.

The high significance of the hearth diameter probably is not due entirely to larger furnaces being more efficient than smaller furnaces, ceteris paribus. Unfortunately, the interpretation of the parameter value of the hearth diameter is clouded by the fact that the larger furnaces are also newer, possess better ancillary equipment, are likely to be operated by more able managers, and are likely to be operated for a full year. In other words, there is strong correlation between hearth diameter and the period of operation; and it is too much to expect that the period of operation adjusts for all missing variables in equation 6.6, while the hearth diameter accounts for only differences in blast furnace size.

To the extent that blast furnace size does make a difference in the coke rate, some of the biases noted in the analysis of the industry level can be explained. In Table 7.4, it is estimated that the weighted,[*] average hearth diameter of all U.S. blast furnaces increased by 3.6 feet from 1950 to 1970. According to equation 6.6, this would cause a reduction of 0.028 in the coke rate.

It should be noted that the parameter values in equation 6.6 comparable with those in the industry level equation 6.5 changed in accordance with expectations. The sinter and burden rate (100/Fe%) parameters were increased slightly in equation 6.6, and the pellet and scrap parameters were reduced greatly. The ranking of their relative effectiveness was preserved, however.

---

[*]The data are weighted by the relative daily production capability of each hearth diameter.

Substituting Other Energy for Coke

Blast Temperature

As discussed earlier in this chapter, the possibility of increasing the temperature of the air blast, all other things being equal, is limited. A blast furnace temperature that is too low or too high can affect adversely the reactions inside the furnace, reducing product quality and possibly even raising the coke rate. For this reason, it is difficult to deduce experimentally the effect of increased temperature on the coke rate. There is always the danger that changes in the coke rate in response to large variations in the blast temperature include the effect of diminishing returns to either the iron-bearing materials or the blast temperature.

The estimates reported in the technical literature on this subject reflect this experimental problem. On the basis of direct experimental observation, T. L. Joseph reported that 100 degrees Fahrenheit is equivalent to 35 pounds of coke per ton of iron; but W. H. Collison reported this equivalence to be 100 degrees Fahrenheit for 25 pounds of coke. [5] Based on a multiple regression analysis of the effect of experimental variations in all furnace inputs on the coke rate, R. V. Flint reported the rate of substitution to be 100 degrees Fahrenheit for 20 pounds of coke; and L. Von Bogdandy, Lange and Heinrich employed an incompletely specified regression equation to analyze operating results on a number of furnaces in various countries to estimate the tradeoff to be 100 degrees Fahrenheit for 10 pounds of coke. [6] The Flint study appears to have been designed more carefully than the others, and his substitution formula will be used in this study.

It is possible that the average blast temperature rose by 400 degrees Fahrenheit from 1950 to 1970. This would have lowered the coke rate by 0.040 (that is, 80 pounds), according to Flint.

Fuel Injection

The first commercial fuel injection system was installed by Colorado Fuel and Iron in their Pueblo works in June 1959. Their system and most of the injection systems installed in the industry during the following seven years used natural gas. In general, the kinds of fuel injectants found to be economically attractive by some firms during the 1960s were natural gas, fuel oil, coal tar, coal dust, and fuel oil-coal dust slurries. Since the rate of substitution of these fossil fuels for coke is almost the same, irrespective of type, attention will be concentrated on natural gas.

The rate of injection typically is measured as a proportion of the air blast. For example, a 1 percent natural gas injection rate means that 1 percent of the air blast is natural gas. Under the prevailing operating practices of most furnace managers in the early 1960s, a 1 percent injection rate was equivalent to 27 pounds of natural gas for each ton of pig iron produced. [7] This equivalence is still approximately correct (as of 1970).

During the 1960s the following was reported:

(1) Each 1 percent increase in the injection rate decreased the coke rate by 60 pounds. Thus, each 45 pounds of natural gas injected saved 100 pounds of coke. Since the natural gas averaged 24,100 BTUs per pound, and coke averaged 15,060 BTUs per pound, it should have been expected that each 45 pounds of natural gas would have saved only 72 pounds of coke. The extra 20-pound saving may have been due to the accompanying increase in the hot blast temperature.

(2) To maintain good working performance in the blast furnace, it was necessary to increase the temperature of the hot blast by 100 degrees Fahrenheit ($\pm 30°F$) for each 1 percent increase in the natural gas injection rate. This is because the breakdown of natural gas (CH_4) into the reducing gases hydrogen (H_2) and carbon monoxide (CO) occurs almost immediately in the tuyere zone (immediately above the furnace hearth), and these reactions are endothermic (heat absorbing) in nature. Furthermore, relatively cold fuel (natural gas) is being substituted for hot carbon units descending the blast furnace stack.

(3) The creation of extra units of energy (CO and H_2) in the top gas at least partly "paid for" the increased energy requirement in the hot blast stoves.

(4) The practical ceiling on fuel injection (dictated by technology rather than by economics was a $7\frac{1}{2}$ percent rate. If practiced at this rate, fuel injection would lower the coke rate by 0.225 tons.

Steam injection also was found to yield favorable results during the last two decades. Furnace operators discovered during the 1950s that the regularity of the furnace performance (and hence the rate of production) was improved greatly when steam was added to the air blast to control humidity inside the furnace. Also, attempts to use higher blast temperatures to save coke were found to be possible on a much larger scale when steam was added to the air blast to limit the increment of heat inside the furnace and, at the same time, to reduce the total heat requirement by causing a greater portion of the iron oxides to be reduced indirectly. (Each 1 percent increase in the steam injection rate increases the proportion of reducing gases in the bosh area by 1.1 percent, adding 0.3 percent carbon monoxide and 0.8 percent hydrogen.)

If steam injection is used in the absence of an increase in blast temperature, the coke rate would increase by about 11 pounds for each 1 percent steam injection increase. But in conjunction with a 100 degrees Fahrenheit increase in the hot blast, each 1 percent increase in steam injection lowers the coke rate from 10 to 13 pounds. If carried

TABLE 6.5

Blast Furnace Consumption of Natural Gas, Fuel Oil, and Tar and Pitch
and Coke Equivalents

	1963	1964	1965	1966	1967	1968	1969	1970
Natural Gas								
Volume (ft^3 × 10^9)	27.3	38.8	46.6	51.5	44.3	46.5	44.4	44.5
Equivalent weight								
(Net tons × 10^3)	552	776	933	1030	885	930	887	889
Energy Equivalent								
(BTU × 10^{12})[a]	27.3	38.8	46.6	51.5	44.3	46.5	44.4	44.5
Fuel Oil								
Volume (gal. × 10^6)	33.8	46.2	53.1	53.2	67.2	80.8	116.8	146.0
Equivalent Weight								
(Net Tons × 10^3)	135	185	213	213	269	323	467	584
Energy Equivalent								
(BTU × 10^{12})[b]	5.4	7.4	8.5	8.5	10.8	12.9	18.7	23.4
Tar and Pitch								
Volume (gal. × 10^6)	n.a.[e]	n.a.	n.a.	n.a.	20.9	22.9	43.4	42.5
Equivalent Weight								
(Net tons × 10^3)	—	—	—	—	84	92	173	170
Energy Equivalent								
(BTU × 10^{12})[b]	—	—	—	—	3.3	3.6	6.9	6.8
Total								
Weight (Net tons × 10^3)	688	961	1,145	1,243	1,238	1,345	1,528	1,644
Energy Equivalent								
(BTU × 10^{12})	32.7	46.2	55.1	60.0	58.4	63.0	70.0	74.7
Consumption: Tons per								
Ton of Pig Iron								
Actual	0.010	0.011	0.013	0.014	0.014	0.015	0.016	0.018
Coke Equivalent								
BTU Basis[c]	0.015	0.018	0.021	0.022	0.022	0.023	0.024	0.027
Total Basis[d]	0.021	0.025	0.029	0.030	0.031	0.033	0.035	0.040

[a]Calculated, assuming 25,000 BTUs per pound of natural gas.
[b]Calculated, assuming 20,000 BTUs per pound of fuel oil.
[c]Calculated, assuming 15,000 BTUs per pound of coke.
[d]Calculated, assuming that 2.2 pounds of coke are replaced by every pound of injected fuel because the blast temperature is increased 100 degrees F for each 1 percent injection rate.
[e]Data not available.

Sources: American Iron and Steel Institute, Annual Statistical Report (New York, AISI, various issues); Handbook of Engineering and Data (London: E & FN Span, 1951).

out at its practical maximum level ($7\frac{1}{2}$ percent), steam injection would lower the coke rate by about 0.045 tons.

If fuel injection is substituted for steam injection, the coke rate saving is only 80 percent of the values reported above[8] because the coke saving attributable to steam injection is lost. Thus, even if complete and accurate data on the practice of fuel injection in the United States were available, it would not be possible to predict accurately the effect on the industry coke rate without additional data on the practice of steam injection. The American Iron and Steel Institute began reporting data on the blast furnace industry's consumption of noncoke fuels in 1963. (It does not, however, report data on blast temperatures or steam injection.) The reported fuels include natural gas, fuel oil, coal tar, and pitch, but exclude coal dust. Table 6.5 reports the volume of consumption of these fuels in their common units of measurement, the equivalent weights in net tons and BTUs. Assuming that the total con- sumption of these materials was for fuel injection (because part could have been used for heating stoves or powering turboblowers), the last column of Table 6.5 indicates the coke savings under the additional assumption that the rising level of fuel injection did not displace steam injection. According to Table 6.5, fuel injection spread rapidly from 1959 to 1965, at which time the consumption per ton of iron was equiv- alent to 0.013 tons, or 26 pounds, that is, a 1 percent injection rate. The increases continued from 1965 to 1969, but at a slower pace. In 1970, the rate of increase jumped again to 36 pounds per net ton of iron, equivalent to a 1.4 percent injection rate. Using the data supplied by Collison[9] and others, the coke rate should have been reduced by 0.021 tons by 1963 and 0.040 tons by 1970. These are fairly modest figures when one considers that it is possible to operate blast furnaces at injection rates of 7.5 percent with proportionate savings in coke rates. The reduction and leveling-off of natural gas consumption from 1966 to 1970 and the relatively rapid increase in the consumption of fuel oil and tar and pitch from 1967 to 1970 indicate that use of these materials as coke substitutes is sensitive to their prices and available supplies. The diffusion of fuel injection may have been restricted by the latter.

MODIFYING THE STATISTICAL ANALYSIS OF INDUSTRY LEVEL DATA:
1950-70 SUMMARY

The initial analysis of the determinants of the coke rate—undertaken by employing regression analysis of industry level data to estimate equation 6.5—now can be improved by incorporating the estimated effect of increases in blast furnace size (equation 6.6) and the reported effect of fuel injection (Table 6.5). The effect of increased blast temperature (that is, increases in temperature not accompanying increased rates of fuel injection) was not considered in the modification of equation 6.5

111

TABLE 6.6

Changes in the Coke Rate, Average Blast Furnace Size, Rate of Fuel Injection, Natural Iron Content of the Iron-Bearing Materials, and Composition of the Iron-Bearing Materials During Selected Intervals from 1949 to 1969

Period	Coke Rate	Average Blast Furnace Size (Hearth Diameter Feet)	Rate of Fuel Injection (Coke Rate Equivalent)	Natural Iron Content (%)	Composition of the Iron Materials			
					Calcined Flux (%)	Scrap (%)	Pellets (%)	Sinter (%)
1949-54	-0.062	1.0	0	1.7	0	2.3	0	7.6
1954-59	-0.088	0.0	0	3.9	1.4	-2.5	8.1	12.3
1959-62	-0.095	0.9	-0.015	1.3	1.2	-0.3	5.9	10.0
1962-64	-0.035	0.5	-0.010	1.1	-0.2	-0.1	7.0	-5.1
1964-69	-0.030	1.2	-0.010	0	0.4	-2.1	16.0	-7.5
1949-69	-0.310	3.6	-0.035	8.0	2.8	-2.7	37.0	17.3

Note: Five-year intervals, except for 1959-64, which was divided into two intervals to pick up the large intraperiod variance in the composition of the iron-bearing materials.

Sources: Coke rate, Table 6.1; average blast furnace size, Chapter 7; fuel injection, Table 6.5; burden statistics, Table 6.2.

TABLE 6.7

Decreases in the Coke Rate Not Explained by Changes in Average Blast Furnace Size, Rate of Fuel Injection, Natural Iron Content, and the Volume of Sinter Flux from 1949 to 1969

	1949-54	1954-59	1959-62	1962-64	1964-69	1949-69
Total Decrease in the Coke Rate	-0.062	-0.088	-0.095	-0.035	-0.030	-0.310
Unexplained Decrease in the Coke Rate[a]	-0.034	-0.024	-0.042	-0.12	-0.006	-0.118

[a]The unexplained decrease is due to the effects of changes in the composition of the iron-bearing materials (and accompanying increases in the hot blast temperature) reported in Table 6.6.

Sources: Total decrease, Table 6.1; explained decrease average furnace size, equation 6.6; fuel injection, Table 6.5; natural iron content, equation 6.5; volume of sinter flux, equation 6.5; fuel injection, Table 6.6; The Making, Shaping and Treating of Steel (Pittsburgh: U.S. Steel Corp., 1964), p. 191.

because of the lack of data at the industry level on this input. This is not a serious omission, as previously explained, because the development of agglomerates paved the way for increases in blast temperatures.

The modification of the parameter values in equation 6.5 was undertaken by considering two- to five-year changes in all pertinent variables see Table 6.6); removing the effect on the coke rate of increases in the natural iron content, rate of fuel injection, average blast furnace size, and the relative volume of sinter flux; and explaining the residual variance in the coke rate by changes in the proportions of sinter, pellets, and scrap in the iron-bearing burden (see Table 6.7). It was assumed that the effects of pellets and scrap on the coke rate were constant in the 1950-70 period and that the effectiveness of sinter in reducing the coke rate improved (because it was documented in Chapter 5 that the quality of sinter improved during this period). By applying the further constraints that scrap was more effective than pellets and pellets more effective than sinter in reducing the coke rate (as predicted in equations 6.5 and 6.6), it was possible to pin down the slope parameter values of the proportions of scrap, sinter, and pellets with reasonable accuracy by assuming the coke rate is a linear function of these proportions.

The parameter values were estimated from the following functional form:

$$\Delta CR = b_1 \left(\Delta \frac{Scrap \%}{100}\right) + b_2 \left(\Delta \frac{pellet \%}{100}\right) + b_3, t \left(\Delta \frac{sinter \%}{100}\right)$$

where ΔCR is the unexplained decrease in the coke rate reported in Table 6.7, b_1 and b_2 are the slope parameters of scrap and pellets, and b_3, t (for t = 1949-54, 1954-59, 1959-64, and 1964-69) is the slope parameter value of sinter. The only parameter values of scrap and pellets fulfilling the constraints noted above were in the neighborhood of -1.000 and -0.300, respectively. These values imply the slope parameter values of sinter were:[*]

- 0.145 in 1949-54,
- 0.200 in 1954-59,
- 0.215 in 1959-64, and
- 0.285 in 1964-69.

In contrast with previous findings based on industry level data, the scrap parameter has approximately the same value as assigned to it in the earlier analyses of the 1900-55 and 1950-70 periods (see equations 6.1d and 6.5); the slope parameter of pellets has been reduced in absolute value by exactly one-third (see equation 6.5);[†] and the sinter

---

[*]The values for 1959-62 and 1962-64 were averaged.

[†]The slope parameter of pellets is still greater in absolute value than the value estimated from company X's operating data in equation 6.6 (-0.230). This is reassuring given the fact the pellets consumed

parameter has been changed in value from $-(0.123 + 0.020d)$ to anywhere in the range from $-(0.145 + 0.006d)$ to $-(0.285 + 0.006d)$, where d is the calcined flux content of the sinter. [*]

These findings indicate that the effectiveness of unfluxed sinter in reducing the coke rate doubled from 1949-54 to 1964-69 and that the difference in the effectiveness of pellets and sinter, in this respect, disappeared during the late 1960s.

---

by company X were of below average quality. (See footnote on p. 145.)

[*] d is expressed in percentage terms. The slope parameter values of d have been adjusted downward by 0.007 (in absolute value) from the values presented earlier to adjust for the negative influence of sinter flux on the natural iron content of the iron-bearing burden.

NOTES

1. The technical information in this section was summarized from The Making, Shaping and Treating of Steel (Pittsburgh: Carnegie Steel Company, 1920), pp. 159-71.

2. The Making, Shaping and Treating of Steel (Pittsburgh: U. S. Steel Corp., 1964), p. 191.

3. Ibid.

4. See, for example, Y. Doi and K. Kasai, "Blast Furnace Practice with Self-Fluxing Sinter Burden," Journal of Metals 11, no. 11 (November 1959): 755-59.

5. T. L. Joseph, "The Potential and Limitations of High Blast Temperatures," Blast Furnace and Steel Plant 49, nos. 3 and 4 (March and April 1961): 239-46 and 324-28. W. H. Collison, "Natural Gas Injection at Great Lakes Steel Blast Furnaces," Iron and Steel Engineer 39, no. 4 (April 1962): 73-81.

6. R. V. Flint, "Effects of Burden Materials and Practices on Blast Furnace Coke Rate," Blast Furnace and Steel Plant 50, no. 1 (January 1962): 47-58; L. von Bogdandy, G. Lange, and P. Heinrich, "Entwicklungsaussichten und Grenzen des Hochofens" ("Prospects of Development and Limits of the Blast Furnace"), Stahl und Eisen 88, no. 22 (October 1968): 1177-1188.

7. For example, see N. B. Melcher, "Bureau of Mines Use of Natural Gas in an Experimental Blast Furnace," AIME Blast Furnace, Coke Oven and Raw Materials Proceedings 18 (1959), 69-74; W. H. Collison, op. cit.; T. L. Joseph, op. cit.

8. J. Michard, "Etude Theorique de l'Injection de Fuel a Temperature de Vent Constante," Journees Internationales de Siderurgie, Luxembourg 1-4 (October 1962).

9. W. H. Collison, op. cit.

7

In this chapter, efforts will be made to estimate the relative contributions of blast furnace size, improved materials, and other kinds of technological improvements to blast furnace productivity (which is usually measured in terms of tons of pig iron output per blast furnace day or year). Because only the data supplied by company X cover the contributions of all these factors, the 1950-70 subperiod will be analyzed first. The results of this analysis will then by applied to the industry level data to isolate the contributions of size and materials improvements in the 1950-70 period. Next, average operating practice in the United States will be compared with best-practice operations in the international sector. Finally, attention will be focused on sources of blast furnace productivity during the 1900-50 period. Figure 7.1 indicates the magnitude of the changes in productivity from 1900 to 1970.

THE 1950-70 PERIOD: TECHNICAL BACKGROUND

Quality of Materials, Gas Flow, and Blast Furnace Top Design

The increases in the natural iron content of the iron-bearing materials estimated in Chapter 5 were achieved by reductions in the proportions of water and impurities. These reductions lowered, in turn, the flux rate and coke rate. These reductions in weight were significant and should have contributed to an approximately proportionate rise in blast furnace productivity. Even more significant, however, were the improvements in the physical quality of the iron-bearing materials. Sinter and pellets are considerably more uniform in size than ores. This is particularly true of pellets.[1] This great

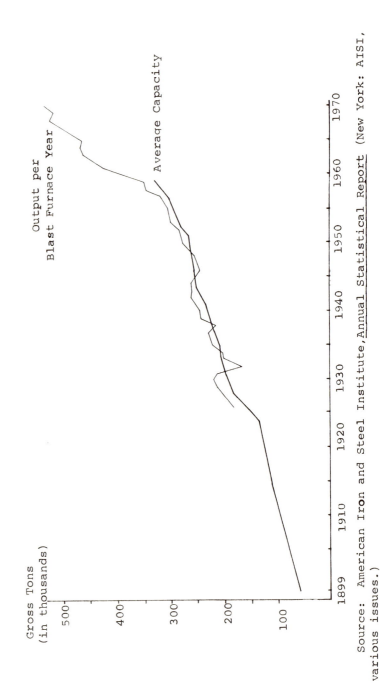

FIGURE 7.1

Average Capacity and Output per Blast Furnace Year—
Coke-Burning Blast Furnaces

Source: American Iron and Steel Institute, Annual Statistical Report (New York: AISI, various issues.)

uniformity of size increases the voidage (the empty spaces) in the stack of the blast furnace. The voidage can range from very low values, for example, 10 to 50 percent when the solid materials are perfectly uniform in size. There is a strong positive correlation between voidage and blast furnace productivity; one experiment showed that as the proportion of sinter particles less than three-eights inch in diameter were decreased from 35 percent (the normal level in 1960) to 5 percent, the production rate of the furnace rose by 32 percent. [2] Another showed that replacing 80 percent of an unscreened ore burden with pellets increased production by 75 percent. [3]

The average particle size is important too. A large size allows high blast rates to be used (because the weight is high relative to surface area), and a small particle size speeds up the reduction and melting of the iron-bearing materials (because the surface area is high relative to the volume) but limits the blast rate that can be used. Since increases in both the blast rate and the speed of reduction and melting raise blast furnace productivity, there is a theoretically optimal particle size, believed to be three-eights inch in diameter. [4]

The blast rate and the size-related characteristics of the materials determine the speed at which the reducing gases pass through the furnace and out through the top. A fast flow tends to waste reducing gases because they have little time to react with the iron oxides. Smaller particles reduce the speed of gas flow, partly alleviating this problem. Another solution is to install pressurized tops. The higher the pressure rating of the top, the slower the gas flow and, hence, the greater the blast rate that can be utilized. However, this source of additional productivity entails costs in addition to the investment in the pressurized top, because the fuel consumption of the blowing engines rises as the blast pressure is increased.

Measures of Blast Furnace Size

Among the available measures of blast furnace size that may be used as direct indicators of furnace output are bosh diameter, hearth diameter, * hearth area, and working volume (the area inside the furnace occupied by the solid and liquid materials). During the 1950s, all but the first of these measures of size were used by operating specialists to indicate the potential practical capacity of furnaces; but during the 1960s, only the last two were considered valid indicators of blast furnace capacity.

---

*During this period, the bosh diameter was from two and one-half to three and one-half feet wider than the hearth diameter. Hence there is little ground for choosing between these two measures.

The hearth diameter was the key size variable in a formula developed by Owen Rice in the 1930s to explain blast furnace productivity for furnaces operating primarily on natural ores and ore concentrates. Rice claimed that blast furnace productivity was proportionate to hearth diameter, for hearth diameters in excess of $13\frac{1}{2}$ feet, given the driving rate (the blast rate) and the coke rate.[*] His formula was based on the ability of the hot blast to penetrate the materials in the furnace directly opposite the tuyeres, which he claimed was an almost constant six feet. Hence, Rice's measure of productivity was based on the area of a ring that was a linear function of the hearth diameter.

During the late 1950s, the development of agglomerates with physical and chemical qualities considerably superior to natural ores and ore concentrates increased the ability of the air blast to penetrate the materials opposite the tuyeres and, as indicated in the last chapter, decreased coke rates, leading to a major reduction in the value of Rice's formula. Blast furnace capacity ratings were then more commonly based on the hearth area or working volume of the furnace. The two measures are interchangeable because almost all furnaces have a working volume to hearth area ratio of 75 to 85 feet.

STATISTICAL ANALYSIS OF COMPANY X'S DATA: 1950-70

The data supplied by company X for selected years in the 1953-70 period were grouped (as in Chapter 6) into 149 observations on all blast furnaces making basic, Bessemer, foundry, and malleable pig iron. A

---

[*]An example of Rice's formula is given in the Anglo-American Council on Productivity Productivity Team Report, Iron and Steel (London: Anglo-American Council on Productivity, 1962), p. 27. Blast Furnace Productivity = A $\times$ R $\times$ F $\times$ 52 weeks, where

A = 6π(Hearth Diameter - 7.5) square feet
R = driving rate = tons of coke burned per week per square foot of A, and
F = the inverse of the coke rate = tons of iron produced per ton of coke burned.

In a 1945 article, "Three Blast Furnace Questions," Blast Furnace and Steel Plant 33, no. 12 (December 1945): 1523-28, Rice claimed that a normal value of R would be 22 tons/wk/A. If the blast furnace had a hearth diameter of 28 feet, then A would be 386.6 feet. If the coke rate were 0.800, then the yearly rated capacity of the furnace at normal operations would be 386.6 X 22 X (1/0.800) x 52 weeks = 550,000 net tons.

log-linear model was first fitted to these data to estimate a value for the exponent of the hearth diameter variable. The value obtained was 1.67 [with a 99 percent confidence interval of (1.42, 1.92)], indicating that, after adjusting for the effects of agglomerates and iron content, blast furnace productivity is not determined as accurately by the Rice ring as it is by the hearth area. Subsequently, a linear model was tested with these data, first using the simple hearth diameter variable, then using the squared hearth diameter as an independent variable. The R^2 was higher by 5 percent when the squared hearth diameter was employed, confirming the log-linear results. Thus, it was decided to employ the squared hearth diameter as the size variable in the regression analysis.*

The major problem was to decide what the proper specification of the productivity equation should be, given the following variables:

(1) hearth diameter;
(2) average natural iron content of the iron-bearing materials;
(3) the percentage breakdown of the iron-bearing materials into scrap, ore, pellets and sinter; and
(4) the period of operation of the blast furnace.†

It was assumed that

(1) an improvement in the burden composition (that is, more agglomerates) should have the same proportional effect on output, regardless of hearth diameter; and
(2) an increase in the natural iron content of the iron-bearing burden, given the burden composition, should have a directly proportional effect on blast furnace output, regardless of hearth diameter.‡

---

*The computer programs available at the time of the analysis permitted the use of only a whole-number valued exponent.
†Recall that

    OP = 1 means a period of operation of 10 days to 1 month,
       = 2 means a period of operation of 1-2 months,
       = 3 means a period of operation of 3-6 months,
       = 4 means a period of operation of 6-10 months, and
       = 5 means a period of operation of 10 months to 1 year.

‡That is, if the natural iron content rises by 10 percent from 51.5 to 56.65, the output rises by 10 percent as well. This can be reasoned from the fact that the unit throughput of iron-bearing materials is reduced by 10 percent, the flux rate will diminish by at least 10 percent and the coke rate will diminish by at least 10 percent. Attempts at using the iron content as a completely independent variable resulted in a smaller estimated effect.

Accordingly, the basic model chosen was this form:

$$\frac{QR}{Fe\%} = A + B_1 \ (Scrap\% \times HD^2) + B_2 \ (Sinter\% \times HD^2)$$

$$+ B^3 \ (Pellet\% \times HD^2) + B_4 \ (OP \times HD^2)$$

where HD^2 is the squared hearth diameter,
OP is the period of operation,
QR is output (in net tons) per blast furnace day, and
Fe% is the average natural iron content of the iron-bearing
materials.

After this model was fitted, it was rearranged to the following form:

$$QR = 0.02 \ (Fe\%) \ \{50A + 0.002 \ HD^2 \ [25,000 \ B_1 \ (Scrap\%)$$

$$+ 25,000 \ B_2 \ (Sinter\%) + 25,000 \ B_3 \ (Pellet\%) + 25,000 \ B_4 \ (OP]\}$$

The company X data were weighted by the number of days of operation for each observation on output per blast furnace day.* The results
are expressed in the following equation: †

$$\hat{QR} = 0.02 \ (Fe\%) \ \{241 + 0.002 \ (HD)^2 \ [-0.08 \ (Scrap\%)$$

$$R^2 = 0.83 + 2.48 \ (Sinter\%) + 5.05 \ (Pellet\%) + 113 \ (OP)\} \qquad (7.1)$$

The "t" values for the weighted regression are inflated considerably
because the weights were applied by multiple counting of some observations, effectively transforming the 149 observations to 1,008 observations. In addition, the "t" values are of limited use in testing the
differences between the parameters of sinter and pellets, and the significance of the scrap parameter and the period of operation parameter,
because all of these variables are mixed with the hearth diameter
variable. Accordingly, tests of significance were applied to the parameter values of the unweighted regression, with the $(HD)^2$ variable
specified as a separate variable. (All parameter values in the
unweighted regression were in the same relative proportions.)

---

*Since the average productivity relationships are designed so the
results of the company X study can be applied to the industry level data,
the regression results will be presented in their weighted form.
†In terms of the pure statistics involved, the R^2 would be higher in
each equation if the $(HD)^2$ variable were separated from the other variables. It is interesting to note that the R^2 is only 0.3 percent higher
than in equation 7.1 when the $(HD)^2$ is separated from the other variables. In this more general form, all other parameter values are in the
same relative proportions.

The scrap parameter value was not found to be significantly different from zero. The sinter parameter value was found to be significantly different from zero at the 85 percent level of confidence. All other parameter values were found to be highly significant. The differential effectiveness of sinter and pellets was found to be highly significant. Notice that the findings concerning the effectiveness of sinter versus pellets completely parallel the coke rate findings in Chapter 6.

The relatively large constant term (241) confirms the log-linear estimation that indicated that the Rice ring theory was partly correct, because the presence of this constant term indicates that output per blast furnace day in the blast furnaces of company X was less than proportionate to the hearth area (but more than proportionate to the Rice ring). If equation 7.1 were refitted with "hearth diameter" included as an additional explanatory variable (that is, in addition to HD^2), the value of the constant term probably would be reduced greatly.

Equation 7.1 can be used to predict output per blast furnace day for various size blast furnaces over a fairly large range of burden compositions. The ranges covered by the mean value plus or minus one standard deviation of each variable in equation 7.1 were

- hearth diameter, 20-28 feet;
- scrap, 0-22 percent;
- sinter, 1-39 percent;
- pellets, 2-56 percent; and
- period of operation, 3.4 to 5 (about five months to one year).

Of course, some observations fell outside these ranges. For example, the upper bounds of the sinter and pellet ranges were increasing over time.

Tables 7.1 and 7.2 report the production rates associated with various combinations of furnace size and burden composition. From the results reported in these tables, the following major points can be deduced:

(1) Increasing blast furnace size and improving the quality of the iron-bearing materials are two important ways to increase blast furnace productivity. For example, based on equation 7.1, if one wished a 60 percent increase in the productivity of a blast furnace with a hearth diameter of 21 feet that had been operating on a natural ore burden with natural iron content of 51.5 percent, it could be achieved either by rebuilding the blast furnace to a hearth diameter of 29 feet or by changing the ore burden to a sinter burden with a natural iron content of 63 percent.

(2) Changing from 100 percent ore with a natural iron content of 51.5 percent to 100 percent pellets doubles the productivity of a blast furnace of given size. Since the quality of company X's pellets is below average, this would represent a conservative estimate.

TABLE 7.1

Predicted Values of Output per Blast Furnace Day for Variations in the Composition
of the Iron-Bearing Materials and Squared Hearth Diameter

Examples	Burden Composition (in percent)				Predicted Output per Blast Furnace Day in Net Tons[a]		
	Ore	Sinter	Pellet	Fe[b]	HD = 21 Feet	HD = 25 Feet	HD = 29 Feet
(1)	100	0	0	51.5	760	975	1,225
(2)	100	0	0	63.0	930	1,195	1,500
(3)	50	50	0	63.0	1,070	1,390	1,764
(4)	0	100	0	63.0	1,205	1,585	2,025
(5)	25	50	25	63.0	1,210	1,590	2,030
(6)	50	0	50	63.0	1,215	1,595	2,035
(7)	0	50	50	63.0	1,350	1,785	2,300
(8)	0	0	100	63.0	1,495	1,990	2,570

[a]Based on equation 7.1, assuming in each case that the furnace is operated for the entire year (OP = 5), and that no scrap is charged.

[b]Except in the first burden composition, high grade ore has been assumed. The iron percentage of sinter, pellets, and high grade ore is 63 percent.

Note: Calculated from equation 7.1.

TABLE 7.2

Index Numbers for Predicted Values of Output
per Blast Furnace Day in Table 7.1

| | Across Burden Compositions | | |
	HD = 21 feet	HD = 25 feet	HD = 29 feet
Burden Composition			
(1) 100 percent ore[a]	100	100	100
(2) 100 percent ore[b]	122	122	122
(4) 100 percent sinter	159	163	165
(8) 100 percent pellets	197	204	210

| | Across Hearth Diameters | | | |
	Burden 1 — 100 percent ore[a]	Burden 2 — 100 percent ore[b]	Burden 4 — 100 percent sinter	Burden 8– 100 percent pellets
Hearth Diameter				
HD = 21 feet	100	100	100	100
HD = 25 feet	128	128	132	133
HD = 29 feet	161	161	168	172

[a]Low grade ore (Fe% = 51.5).
[b]High grade ore (Fe% = 63.0).
Source: Based on equation 7.1, assuming in each case that the furnace is operated for the entire year (OP = 5) and that no scrap is charged.

(3) The productivity of a blast furnace is related more closely to the hearth area than to the hearth diameter. For example, a change in the hearth diameter from 21 feet to 29 feet represents a 38 percent increase, but the rate of production for company X was found to increase by 61 to 68 percent for such an increase in the hearth diameter. If output had been directly proportional to hearth area, it would have increased by 91 percent.

(4) The period of operation was found to be important in explaining a furnace's daily output. This variable can be thought of as a proxy for furnaces producing only at times of peak demand, furnaces producing special types of pig iron (such as foundry and malleable), and for furnaces in older plants, since all of these conditions can be associated with short campaign times on a blast furnace.

At this point equation 7.1 will be applied to the industry level data
on burden composition and output per blast furnace day to discover if a
fit can be obtained with acceptable values for the key variables missing
from the industry-level data: the weighted average hearth diameter and
the weighted average period of operation per blast furnace. If these
values are acceptable, it will then be possible to partition the growth
in output per blast furnace day in the industry between increases in
furnace size and improvements in materials.

For the data set of 149 observations, the arithmetic average hearth
diameter was 22.5 feet. The geometric average was 22.9 feet, and the
weighted geometric average was 24.25 feet. For the industry as a whole
in 1966, the arithmetic average hearth diameter was 23.7 feet, [5] and the
geometric average was 24.1 feet. An educated guess would put the
weighted geometric average hearth diameter at 25.5 feet. Substituting
this value and the 1966 values for scrap, sinter, pellets, iron content, [6]
and output per blast furnace day into equation 7.1 yields the value of
4.41 for the period of operation variable, which compares favorably to
the 4.38 value obtained for company X's operations in the entire 1953-70
period. Applying equation 7.1 to the industry level data appears to be
well within the realm of acceptability. From the basic model we obtain
the equation below.

$$QR = 0.02(Fe\%) \quad \{A' + 0.002 \, (HD^2) \, [B_1' \, (Scrap\%) + B_2' \, (Sinter\%)$$
$$+ B_3' \, (Pellet\%) + B_4' \, (OP)]\}$$

where $\quad A' = 50A$

and $\quad B_j' = 25,000 \, B_j{}^*$

A change in the value of QR to QR + ΔQR can be expressed as shown.

$$QR + \Delta QR = 0.02(Fe\% + \Delta Fe\%) \, \{A' + 0.002 \, (HD^2 + \Delta HD^2) \, [B_1' \, (Scrap\%$$
$$+ \Delta Scrap) + B_2' (Sinter\% + \Delta Sinter\%) + B_3' \, (Pellets\%$$
$$+ \Delta Pellets\%) + B_4' \, (OP)]\}$$

Expansion and simplification of this equation yields a somewhat
cumbersome form.

---

*This equation is a restatement of the basic model that was used
to determine the structure of the regression equations at the beginning
of this section.

$$\Delta QR = (0.02)(0.002)(Fe\%)[B_1'(Sc\%) + B_2'(Si\%) + B_3'(P\%) + B_4'(OP)](\Delta HD^2)$$
$$+ (0.02)\{A' + 0.002(HD^2)[B_1'(Sc\%) + B_2'(Si\%) + B_3'(P\%) + B_4'(OP)]\}(\Delta Fe\%)$$
$$+ (0.02)(0.002)(Fe\%)(HD^2)B_1'(\Delta Sc\%)$$
$$+ (0.02)(0.002)(Fe\%)(HD^2)B_2'(\Delta Si\%)$$
$$+ (0.02)(0.002)(Fe\%)(HD^2)B_3'(\Delta P\%)$$
$$+ (0.02)(0.002)(Fe\%)(HD^2)B_4'(\Delta OP)$$

- -

$$+ (0.02)(0.002)[B_1'(Sc\%) + B_2'(Si\%) + B_3'(P\%) + B_4'(OP)](\Delta Fe\%)(\Delta HD^2)$$
$$+ (0.02)(0.002)(Fe\%)[B_1'(\Delta Sc\%) + B_2'(\Delta Si\%) + B_3'(\Delta P\%) + B_4'(\Delta OP)](\Delta HD^2)$$
$$+ (0.02)(0.002)(HD^2)[B_1'(\Delta Sc\%) + B_2'(\Delta Si\%) + B_3'(\Delta P\%) + B_4'(\Delta OP)](\Delta Fe\%)$$

- -

$$+ (0.02)(0.002)[B_1'(\Delta Sc\%) + B_2'(\Delta Si\%) + B_3'(\Delta P\%) + B_4'(\Delta OP)](\Delta Fe\%)(\Delta HD^2)$$

The terms down to the first dotted line represent the first order effects. The terms between the first and second dotted lines represent the second order effects, and the last term represents the third order effects. The third order effects can be neglected without noticeably affecting the partitioning of ΔQR into its various components. Ignoring the second order effects will have a more noticeable effect on the partitioning of ΔQR. The seriousness of neglecting the second order effects depends on the time lapse for which ΔQR is defined. For a time lapse of 10 years, the effects will be noticeably large. It was decided to calculate the first order components of ΔQR over intervals of five years from 1950 to 1970. The first order effects accounted for 94.5 percent of the change in ΔQR using this format. The second order effects were ignored, since each component cannot be attributed to the change in one specific variable.

A value of 4.5 was assigned (arbitrarily) to the OP variable, since the only information on this variable comes from company X's data and the industry level data in 1966. Given this value, and given the values for all other variables except the hearth diameter, the average hearth diameter was calculated in each of the years 1950, 1955, 1960, 1965, and 1970. The accuracy of these calculated values depends on the accuracy of the specification of the productivity model and the accuracy of the assumption that the period of operation was approximately 4.5 in each of the years under consideration. One test of the accuracy of these items is to see if the resulting calculated values of the average hearth diameter increase gradually from period to period, as would be expected. Table 7.3 contains the values of the variables in question and reports the calculated values of the weighted, average hearth diameter. The calculated values do appear to be reasonable—certainly well within the realm of possibility.

TABLE 7.3

Calculated Values of the Industry-Level, Weighted Geometric Average Hearth Diameter

Year	Scrap (in percent)	Sinter (in percent)	Pellet (in percent)	Fe[a] (in percent)	OP[b]	Output[c]	Calculated HD[d]
1950	8.7	15.5	0.0	52.55	4.5	848	22.75
1955	10.6	17.3	1.5	53.81	4.5	937	23.75
1960	8.0	42.1	9.7	59.64	4.5	1182	23.80
1965	7.7	37.2	23.7	62.37	4.5	1436	25.15
1970	5.7	30.1	40.3	61.67	4.5	1642	26.35

[a] Calculated from Table 6.1 by adjusting for the percentage of calcined flux in the iron-bearing materials.

[b] Assumed to be 4.5, based on company X data.

[c] Per blast furnace day (net tons).

[d] Calculated, using equation 7.1.

Source: Table 6.1 and Annual Statistical Report (New York: American Iron and Steel Institute, various years).

TABLE 7.4

Sources of Increase in Output per Blast Furnace Day

Period	Total Increase in Output per Blast Furnace (net tons)	Increase in Output Attributable to the Iron-Bearing Burden					Increase in Output Due to Increases in Average Hearth Diameters of U.S. Blast Furnaces	Total Explained Increase
		Physical Properties of the Iron-Bearing Burden			Natural Iron percent of the Burden	Total		
		Scrap	Sinter	Pellets				
1950-55	98	-0.2	4.9	8.2	20.3	33.2	53.4	86.6
1955-60	245	0.3	74.9	50.2	101.5	226.9	2.4	229.3
1960-65	254	0	-16.8	95.6	54.1	132.9	103.0	235.9
1965-70	206	0.3	-27.8	132.2	-16.1	88.6	110.3	198.9
Totals: 1950-70	794	0.4	35.2	286.2	159.8	481.6	269.1	750.7

Source: Equation 7.1.

127

Given the values for the proportions of scrap, sinter, and pellets; their average iron content; the period of operation; and the average hearth diameter; the five-year increase in output per blast furnace day were partitioned into their primary sources. These results are reported in Table 7.4.

To summarize, from 1950 to 1970 changes in the quality of materials accounted for 64 percent of the explained increase in output per blast furnace day, with increases in blast furnace size accounting for the other 36 percent. The materials improvements appear to be tapering off in the 1960s. Increases in the natural iron content accounted for 21 percent of the explained increase in output per blast furnace day, with the improved physical properties of the iron-bearing burden accounting for the other 43 percent. The potential for increases in the natural iron content to improve blast furnace productivity appears to be almost exhausted. Further increases in furnace productivity from materials improvements probably will continue to stem from the substitution of pellets for sinter, sinter for ore, and a gradual increase in the physical quality of sinter.

SOME QUALIFICATIONS

The previous estimates were made as if increases in blast furnace size and improved materials were the only sources of increased blast furnace productivity. Actually, in addition to these sources there were three other significant areas of innovation in blast furnace practice during these two decades: pressurized tops, the substitution of injected fuels for coke, and improved process controls. Accordingly, a brief discussion of the probable effects of these innovations on the above estimates will be undertaken, assuming company X was an equal participant with other firms in adopting these innovations.

High Pressure Tops

Republic Steel pioneered the development of high pressure tops in the late 1940s. Although data on the diffusion of this innovation are poor, it appears that these tops were in general use by the mid-1950s. This rapid diffusion can be attributed to the resulting significant reductions in flue dust and improvements in coke rates and furnace productivity. Republic Steel tended to install pressurized tops on its largest and most recently built (or rebuilt) furnaces first. As of 1970, some of their smaller furnaces still did not have these tops. It may be surmised that other steel companies followed the same pattern. If so, part of the effect of increases in hearth diameter on blast furnace productivity (established in the last section with the use of company X's data) may be attributable to the positive correlation between the use of pressurized tops and size.

Most furnaces with pressurized tops are rated at only 5 to 15 pounds per square inch above gravity (p. s. i. g.). Development of tops with even greater pressure ratings has been slow, despite general agreement that this would lead to improvements in furnace productivity. In 1959, Koppers announced the design of a blast furnace with a top pressure of 40 p. s. i. g., which at that time was more than four times as high as any in operation.[7] In 1969, Meinhausen cited two atmospheres (28 p. s. i. g.) as the best-practice furnace operation at that time.[8] Thus, the Koppers design appears to have been ahead of its time. Perhaps one reason very high pressure furnaces have been slow to appear is that such a furnace must be built entirely anew—with the blast furnace, hot blast stoves, and other parts designed as pressure vessels, with very powerful turboblowers, with special top seals to prevent leakage of the top gas, and with two casthouses—since the productivity of such a furnace would be double that of a blast furnace operating under normal pressure.

Fuel Injection

As noted in the previous chapter, fuel injection was first practiced in 1959, and it spread rapidly enabling all U. S. furnaces in operation to have an average injection rate of 1 percent in 1963.[9] By 1970, the average injection rate was only 1.4 percent. The slow diffusion from 1963 to 1970 is puzzling in light of the widespread claims that, up to an injection rate of 7 percent, this practice not only permits coke savings in excess of the fuels that replace it but also improves blast furnace productivity by
(1) directly decreasing the coke rate;
(2) lowering the proportion of iron oxides that are directly reduced (through contact with the burning coke), thereby decreasing the coke rate further; and
(3) increasing the optimal temperature of the hot blast, thereby speeding up the rate at which the solid materials are melted.

The first two points were documented in the previous chapter. The last point has been noted by a number of engineers who published results of experiments (or initial commercial efforts) with fuel injection. W. H. Collison reported a $1\frac{1}{2}$ percent increase in productivity for each 1 percent increment of fuel injection,[10] although he was not definite about the increase in the hot blast accompanying the fuel injection. T. L. Joseph reported a 3 percent increase in the productivity of the Bureau of Mines experimental blast furnace for each 1 percent increment in fuel injection accompanied by a 100 degrees Fahrenheit increase in

129

the hot blast temperature.[11] The most dramatic productivity increase was reported by Compagnie des Ateliers et Forges de la Loire at their Boucau Works. At a 5 percent fuel oil injection rate with an unspecified increase in the blast temperature, they reported a 23 percent increase in blast furnace productivity.[12] The most commonly experienced increases are in the range of 2-3 percent for each increase of 1 percent in the injection rate, according to J. H. Strassburger and others.[13] However, since the same productivity gains can be obtained from steam injection that are obtained from fuel injection,[14] it is possible that the average level of the less costly steam injection was high enough to retard the diffusion of fuel injection. At the estimated fuel injection rate prevailing in 1970, $4\frac{1}{2}$ percent to $6\frac{1}{2}$ percent of the increase in blast furnace productivity attributed to improvements in materials from 1950 to 1970 in the previous section was actually the result of this practice. It may be argued, however, that improved materials and fuel injection are complementary innovations. Improvements in the former, by increasing the throughput of all materials (including coke) in a furnace of given size and by improving gas-solid contact, make investment in injection equipment more attractive and improve the effect of fuel injection on blast furnace productivity. Finally, the magnitude of the impact of all injected materials on furnace productivity may be considerably higher when the effect of steam injection is added to that of fuel injection. Although data pertaining to the practice of steam injection are not available, it is possible that the steam injection rate was as high as 3 percent in 1970, which would mean that the effect of improved materials on furnace productivity was overstated by 14 percent to 20 percent.

Process Controls

Improved controls provide a number of ways to increase blast furnace productivity. Computer controls adjust the humidity of the blast almost instantaneously, permitting the reactions inside the furnace to occur as smoothly as possible and almost eliminating slips and hangs within the furnace. Given the quality of the blast furnace materials, controls have been established for the optimal charging sequence, including the quantities of the various materials per charge, the frequency of charging, and the sizing of the materials. Rapid qualitative analysis of the molten iron in the hearth of the furnace has been developed. The samples can be removed while the furnace is in operation through a tube that is inserted through one of the tuyeres.

The cumulative impact of these improvements in controls on furnace productivity is not known at the industry level, although it is probably substantial. These improvements could tend to overstate the effects of both increases in hearth diameter and improvements in materials on blast furnace productivity (estimated in previous sections), since all three are positively correlated.

IMPROVEMENTS SINCE 1950 IN U.S. BLAST FURNACE TECHNOLOGY AND THE FUTURE POTENTIAL

One accepted measure of technological improvements in blast furnace operations is to measure the increase in output per blast furnace day per square foot of hearth area. The underlying hypothesis is that this measure would be the same for a large or small furnace if everything else were the same. In a 1970 interview, David Dilley of United States Steel Corporation stated that the current U.S. industry average was three net tons a day per square foot of hearth area. In best-practice operations, Dilley revealed that five net tons was just being reached in U.S., while Japan was attaining six net tons in some of their newest, most sophisticated furnaces. In a 1969 article, Meinhausen estimated that best-practice operations in 1962 could achieve 4.6 net tons without a high pressure top or fuel injection, but with a high percentage of high quality agglomerates, high blast temperature, and (presumably) steam injection.[15] For 1969, Meinhausen claimed that best-practice operations could reach 6.2 net tons per square foot of hearth area, which is in agreement with the statement by Dilley. To achieve this performance, Meinhausen specified that the blast furnace would have to operate at a high top pressure of two atmospheres (28 p.s.i.g.), with a hot blast temperature of 2370 degrees Fahrenheit (achieved by use of external combustion chambers) in conjuction with fuel injection, with a high percentage of high quality agglomerates, and computerized process control. If the top pressure were to be increased to three atmospheres (42 p.s.i.g.), this figure could be improved to 8.2 net tons per square foot of hearth area, which is in agreement with estimates made by Koppers.

The estimates of the weighted average hearth diameter, derived from equation 7.1, can be used to indicate the diffusion of these innovations in average U.S. practice. There is one precaution, however, that must be mentioned. Since the form of equation 7.1 indicates that the productivity of company X's blast furnaces was less than directly proportional to the hearth area, given the burden composition, smaller blast furnaces will be calculated to have a higher output per blast furnace day per square foot of hearth area than larger blast furnaces. For example, if an "average furnace" in 1950 is compared with its average 1970 counterpart, both furnaces operating with an iron-bearing burden that is average for this 20-year period (30 percent sinter, 15 percent pellets, and a natural iron content of 58 percent, the technology coefficient for the smaller furnace is calculated to be 2.62, while the larger furnace's coefficient is 2.44, a relative discrepancy of

7 percent. With this precaution in mind, the technology coefficients for average U.S. practice were calculated to be

- 1950, 2.1;
- 1955, 2.1;
- 1960, 2.65;
- 1965, 2.9; and
- 1970, 3.0.

Notice that 1970 is in exact agreement with Dilley's estimate.

These figures indicate about a 50 percent increase in output per blast furnace day that is attributable to technological change in the materials (and accompanying changes in blast rates, blast temperatures, fuel injection, and process controls) used in blast furnace operations. Given only modest adoption of improvements in high pressure tops by blast furnace managers, it appears reasonable to claim that this factor played a minor role in the increase in output per blast furnace day attributed to technological change.

It may be interesting to note that, if U.S. blast furnaces all operated on 100 percent pellet burdens with a natural iron content of 63 percent, output per blast furnace day would rise from its 1970 level of 1642 net tons to 2077 net tons, assuming that the period of operation remained at 4.5 and the weighted average hearth diameter at 26.35 feet. This would raise the technology coefficient to 3.8 net tons per day per square foot of hearth area, an increase of 25 percent. The 1970 burden composition of 40 percent pellets and 30 percent sinter yielded an iron content that was close to the maximum value that reasonably can be considered attainable (63-64 percent) but represented only 55 percent of the potential improvement in the physical properties of the iron-bearing burden (100 percent pellets). Even if the average period of operation were increased from 4.5 to 5 (a 10 to 12 month minimum campaign period), the technology coefficient would reach only 4.0. This figure is still 15 percent short of Meinhausen's claim that 1962 best-practice operations would yield a technology coefficient of 4.6, implying that U.S. furnaces are not being driven as hard as they could be, and/or that they are not utilizing hot blast temperatures and steam injection to the fullest extent.

THE 1900-50 PERIOD

Equation 7.1 fits operations in this earlier period with surprising accuracy when a simple modification of the equation is made. The modification is the replacement of the constant term (241) with 9.94(HD), where 9.94 is the quotient of 241 divided by the weighted average hearth diameter of company X's furnaces during 1953-70 (which was 24.25 feet), yielding the equation below.

$$\hat{QR} = 0.02(Fe\%) \{9.94(HD) + 0.002(HD^2)[-0.08(scrap\%) + 2.48(sinter\%)$$

$$+ 5.05(pellet\%) + 113(OP)]\}. \qquad (7.2)$$

If equation 7.2, rather than equation 7.1, is used to estimate the weighted average hearth diameter of all utilized U.S. furnaces in 1950 and 1970, the calculated values are 23.0 feet and 26.1 feet, respectively. These estimates are nearly the same as those based on equation 7.1 (which were 22.75 feet and 26.35 feet).

Equation 7.2 will be used for two analytical purposes in the 1900-50 period. First, employing estimates of the average period of operation in the earlier period, it will be used to yield estimates of the weighted average hearth diameter of utilized U.S. furnaces at approximately 10-year intervals. Second, the resulting estimates of the average hearth diameter will be used in conjunction with figures on output per blast furnace day to calculate "output per blast furnace day per square foot of hearth area" to measure technological improvements in average operating practice during this period.

Increases in Average Blast Furnace Size

The average hearth diameter of utilized furnaces in 1908 was calculated directly to be 13.75 feet (see Figure 7.2 for a comparison of furnace sizes in 1908 and 1966).[16] Since output per blast furnace day was approximately 300 net tons in 1908, the average period of operation was 3.6 (that is, slightly in excess of six months).[17] Using 1908 and the 1950-70 period as benchmarks, the average period of operation was estimated to be 3.0 in 1899, 4.0 in 1919, 4.4 in 1929, and 4.3 in 1939.[18] Table 7.5 reports the resulting calculated values of the average hearth diameter in those years and the possible error range.

These hearth diameter estimates partially can be checked by comparing their increases within each interval with the building and scrapping of blast furnaces in the same interval. Specifically, it was hypothesized that the important determinants of the increase in the weighted, average hearth diameter over any period of time are the percentage of the blast furnace stock at the beginning of the period that was scrapped and the percentage of the blast furnace stock at the end of the period that was newly built during that period. To keep the test simple, the percentage of scrapped furnaces was added to the percentage of newly built furnaces to construct a single explanatory variable. Table 7.6 summarizes the result of this test.

A very good fit was obtained (see footnote c of Table 7.6). The unadjusted R^2 indicates that the relative frequency of building and

TABLE 7.5

Estimated Values of the Average Length of Operation and Hearth Diameter
for Industry Operations in Selected Years

Year	Capacity Utilization[a] (percent)	Estimated Value of OP[a]	Estimated Hearth Diameter (feet)	Error in Estimated HD if OP Were Higher or Lower by 1 (feet)
1899	64	3.0	10.0[b]	0.8
1908	44	3.6	13.75	1.1
1919	63	4.0	15.5	1.4
1929	83	4.4	20.75	2.1
1939	63	4.3	22.5	2.2
1950	91	4.5	23.0	2.2

[a]The effect of capacity utilization on OP (the period of operation) was found to be fairly small. It is assumed that by 1929 the industry organization and technological developments in furnace linings had proceeded to the point where potential campaign times were sufficiently above one year to make the average value of OP almost the same as during the 1950s and 1960s.

[b]Application of the Rice ring concept to furnaces less than 13.5 feet in hearth diameter is potentially mis-leading, because the ring becomes the hearth area for such furnaces.

Source: Equation 7.2.

134

TABLE 7.6

Comparison of Increases in the Predicted Average Hearth Diameter
to the Rates of Scrapping and New Building of Blast Furnaces

		Furnaces Scrapped During the Period			Furnaces Built During the Period		Sum	HD Increase	
Year	Number of Coke Blast Furnaces on Dec. 31	Period	Number	As a percentage of Furnaces at the Beginning	Number	As a percentage of Furnaces at the End	$(A + B)$ (in percent)	Predicted[c] (feet)	Previously Estimated by Equation 7.2 (feet)
1899	350[a]								
1908	401	1900–08	53	15	104	26	41	3.85	3.75
1919	421	1909–19	56	14	76	18	32	2.61	1.75
1929	277	1920–29	168	40	24	9	49	4.94	5.25
1939	222	1930–39	59	21	4	2	23	1.38	1.75
1950	250[b]	1940–50	9[b]	4	29[b]	12	16	0.42	0.50

[a]This is an estimate from the 1899 Census of Manufactures (Washington, D.C.: Bureau of the Census).

[b]The indicated net increase of 28 blast furnaces is greater than the 29 new minus the 9 scrapped equals 20 furnaces reported in this table. Of the missing 8 furnaces: 2 were revived during the war years, 1 was built as an experimental blast furnace, 1 was a very small furnace built to produce ferroalloys, and 4 are unaccounted for.

[c]The predictions are based on the fitted equation: $\widehat{\Delta HD}$ (in feet) = $-1.77 + 0.137\,(A + B)$, where A = percentage of scrapped furnaces and B = percentage of newly built furnaces. The (unadjusted R^2 is 0.93 and the "t" value for 0.137 is 1.59, implying this parameter is significant at the 85 percent confidence level.

Source: American Iron and Steel Institute, Annual Statistical Reports (New York: AISI, all issues prior to 1951).

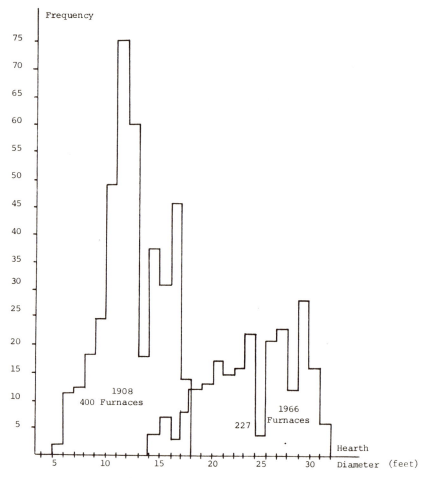

FIGURE 7.2

Distribution of Coke-Fueled Blast
Furnaces by Hearth Diameter in 1908 and 1966

Note: Hearth diameters of furnaces in 1908 were calculated, assuming the reported bosh diameters were six feet greater than the unreported hearth diameters.

Source: Directory of Iron and Steelworks in America (New York: American Iron and Steel Institute, 1908 and 1966).

scrapping within each 10-year interval was capable of explaining 93 percent of the variance in increases in average furnace size. Since rebuilding (which is an important source of blast furnace enlargement) and capacity utilization figures were omitted from the regression, the R^2 is surprisingly high. This high value may be misleading, however, because of the very limited number of observations. The evidence indicates that the major increases in average furnace size occurred in the 1900-08 period (primarily from the large number of big furnaces that were constructed) and in the 1920-29 period (during which 40 percent of the furnaces in existence in 1920 were scrapped).

A Measure of Technological Improvements
in Blast Furnace Operations

Utilizing the estimates of the average hearth diameter derived from equation 7.2 and figures on output per blast furnace day, output per blast furnace day per square foot of hearth area was calculated at (approximately) 10-year intervals from 1908 to 1950 (see column 3 of Table 7.7).* This value remained constant at "2" in 1908, 1919, and 1950 and was slightly less than "2" in 1929 and 1939.

Since equation 7.2 indicates that the Rice ring concept is partially valid (that is, that increases in furnace output are less than proportionate to increases in furnace hearth area), a value for output per blast furnace day per square foot of hearth area was calculated for a furnace of average size in each of these years, under identical operating conditions, to determine a "technological norm" against which the estimated actual values can be contrasted (compare columns 3 and 4 of Table 7.7). Dividing each estimated actual value by the value that would have prevailed under fixed operating conditions (column 5 of Table 7.7) and normalizing those quotients to "1908 = 1.00" yields estimates of improvements in average blast furnace technology that occurred independently of increases in the size of blast furnaces (column 6 of Table 7.7). This approach indicates an improvement of 22 percent in blast furnace productivity between 1908 and 1950 that was not related to increases in size (including technological improvements that were related in a cause and effect manner with increases in furnace size, such as mechanical charging). This increase of 22 percent probably was caused by improvements in the linings and tops of furnaces, better process controls, and higher blast rates and temperatures. Improvements in iron-bearing materials were not a factor in this period.

---

*1899 was omitted in Table 7.7 because the hearth areas of furnaces in that year were—on the average—less than the Rice ring.

TABLE 7.7

Estimates of Improvements in Blast Furnace Technology: 1908–50

Year	Output Per Blast Furnace Day[a] (net tons) (1)	Estimated Hearth Diameter[b] (feet) (2)	Output per Blast Furnace Day Per Square Foot of Hearth Area[c] (3)	Output per Blast Furnace Day per Square Foot of Hearth Area Under Fixed Operating Conditions[d] (4)	Technology Coefficient [(3) ÷ (4)] (5)	Technology Coefficient Relative to 1908 [(5) ÷ 0.88] (6)
1908	299	13.75	2.01	2.28	0.88	1
1919	382	15.5	2.02	2.17	0.93	1.06
1929	655	20.75	1.93	1.94	0.99	1.13
1939	736	22.5	1.85	1.91	0.97	1.10
1950	848	23.0	2.03	1.90	1.07	1.22

[a]1908 and 1919 values are estimates derived from column b of Table 7.8. 1929, 1939, and 1950 values were reported by the American Iron and Steel Institute in their Annual Statistical Report (New York: AISI, 1929, 1939 and 1950).

[b]See Table 7.5.

[c]Column 1 divided by the hearth area calculated from column 2.

[d]The fixed operating conditions were 100 percent ore burden, a natural iron content of 51.5 percent, and a period of operation of 4.5. Using these values and the hearth diameters reported in column 2, a figure for output per blast furnace day was calculated from equation 7.2 and subsequently used to derive the figures in column 4.

Source: Compiled by the author.

THE EFFECT OF FLUCTUATIONS IN DEMAND
ON FURNACE PRODUCTIVITY

In an industry that is strongly subject to general economic conditions, it could be expected that in recessionary periods the member firms could at least cease operations in their least efficient plants and equipment units to compensate for the reduced volume of sales. In the blast furnace industry, the newest and largest furnaces tend also to be the most efficient. It is evident that the rate of operations cannot be lowered significantly on a furnace in blast without adversely affecting its operations.

Thus, to examine the possibility that during periods of low capacity utilization managers operate only their most efficient furnaces, annual data on output per blast furnace year were divided by the average capacity of all furnaces in the 1926-59 period and regressed against two explanatory variables: a quadratic function of capacity utilization and the average growth rate of average furnace capacity for the last two years. One year, three years and four years were also tried. The four-year growth rate gave results distinctly inferior to the other measures.) The data used are reported in Table 7.8. The results of the regression were as shown below (standard errors are in brackets).

$$(\frac{QR}{Ave\ Cap}) = 0.8412 + 0.0284\,(\bar{g}_2) + 0.4290\,(CU_t) - 0.2720\,(CU_t)^2$$

$$R^2 = 0.774 \quad [0.0036] \quad [0.1730] \quad [0.1368] \qquad (7.3)$$

where   QR = output per blast furnace year (in gross tons)

Ave Cap = the average yearly capacity of coke blast furnaces

$\bar{g}_2$ = the average rate of growth in average capacity over the previous two years

CU_t = capacity utilization during the year

All parameters are significant at least at the 95 percent level of confidence. In all regressions of this basic form, the coefficients were of the same approximate magnitude and sign. Equation 7.3 can be rewritten as shown.

$$(\frac{\hat{QR}}{Ave\ Cap}) = 0.028\,(\bar{g}_2 - 2.52) + 0.913\,(1 + 0.470\ CU_t - 0.298\ CU_t^2)$$

$$(7.4)$$

Equation 7.4 informs us that if the average two-year growth rate is 2.52 percent (its average value), the ratio of output per blast furnace year to the average blast furnace capacity will be

TABLE 7.8

Capacity, Output, Capacity Utilization, Number of Coke Blast Furnaces, Average Capacity,
Growth Rates of Average Capacity, and Output per Blast Furnace Year

Year	Dec. 31 Capacity, Gross Tons ×10³	Output, Gross Tons ×10³	Capacity Utilization	Dec. 31 Number of Blast Furnaces	(A) Dec. 31 Average Capacity, Gross Tons ×10³	Rate of Average Capacity Growth 1 Year (percent)	2 Year Average (percent)	(B) Output per Blast Furnace Year, Gross Tons ×10³	Ratio of B to A
1899	(20,900)ᵃ	(13,336)ᵃ	(64%)ᵃ	(350)ᵃ	59.7	—	—	(63.7)ᶜ	(1.067)ᶜ
1908	36,427	15,469	(44%)ᵃ	401	90.8	(4.25)ᵃ	—	(96.0)ᵇ	(1.102)ᵇ
1912	42,563	29,037	(70%)ᵃ	421	101.1	—	—	—	—
1913	—	30,278	71.1	—	—	—	—	—	—
1914	43,757	22,734	(53)ᵃ	409	107.0	—	—	—	—
1915	44,387	29,229	66.8	405	109.6	—	—	—	—
1916	45,229	38,391	86.5	408	110.9	—	—	—	—
1917	47,377	37,521	83.0	420	112.8	—	—	—	—
1918	48,628	37,752	79.7	422	115.2	—	—	—	—
1919	49,628	30,216	62.1	421	117.9	2.34	2.24	(122.9)ᵇ	(1.067)ᵇ
1920	51,125	35,921	74.4	420	121.7	—	—	—	—
1921	51,924	16,304	31.9	422	123.0	—	—	—	—
1922	52,124	26,600	51.2	421	123.8	—	—	—	—
1923	52,146	39,470	75.7	408	127.8	—	—	—	—
1924	52,896	30,662	58.8	401	131.9	—	—	—	—
1925	50,831	35,920	67.9	347	146.5	—	—	—	—
1926	52,110	38,534	75.8	334	156.0	6.48	8.48	180.1	1.229
1927	50,330	35,693	68.5	297	169.5	8.65	7.57	189.4	1.214
1928	51,070	37,259	74.0	286	178.6	5.37	7.01	204.1	1.204
1929	51,490	41,619	81.5	277	185.9	4.09	4.73	210.6	1.179
1930	52,516	30,924	60.1	272	193.1	3.87	3.98	217.4	1.169
1931	51,598	17,912	34.1	265	194.7	0.83	2.35	214.6	1.111
1932	50,314	8,535	16.5	256	196.5	0.92	0.88	168.8	0.867
1933	50,975	12,969	25.8	253	201.5	2.54	1.73	195.1	0.993
1934	50,846	15,651	30.7	251	202.6	0.55	1.55	200.2	0.994
1935	49,778	20,706	40.7	241	206.5	1.92	1.24	216.4	1.068
1936	49,513	30,140	60.5	234	211.6	2.47	2.20	222.5	1.077
1937	50,606	36,049	72.8	233	217.2	2.65	2.56	225.0	1.063

Year	Dec. 31 Capacity, Gross Tons ×10³	Output, Gross Tons ×10³	Capacity Utilization	Dec. 31 Number of Blast Furnaces	(A) Dec. 31 Average Capacity, Gross Tons ×10³	Rate of Average Capacity Growth 1 Year (percent)	Rate of Average Capacity Growth 2 Year Average (percent)	(B) Output per Blast Furnace Year, Gross Tons ×10³	Ratio of B to A
1938	50,199	18,507	36.6	226	222.1	2.26	2.46	213.8	0.984
1939	49,668	31,036	61.8	222	223.7	0.72	1.49	236.6	1.065
1940	51,342	41,059	82.7	223	230.2	2.91	1.81	241.1	1.078
1941	53,828	44,107	95.6	233	231.0	0.35	1.63	238.5	1.079
1942	56,988	52,653	97.8	238	239.4	3.64	2.00	254.9	1.103
1943	60,121	54,230	95.2	245	245.4	2.51	3.08	258.7	1.081
1944	60,045	54,418	90.5	241	249.1	1.51	2.01	259.7	1.088
1945	60,097	47,478	79.1	240	250.4	0.52	1.02	252.0	1.012
1946	58,640	39,981	66.5	232	252.8	0.96	0.74	241.7	0.965
1947	60,178	52,079	88.8	238	252.9	0.03	0.50	248.5	0.983
1948	62,948	53,621	89.1	245	256.9	1.58	0.81	252.3	0.998
1949	63,802	47,690	75.8	246	259.4	0.97	1.28	260.3	1.013
1950	64,707	57,667	90.4	250	258.8	-0.23	0.37	272.6	1.051
1951	65,877	62,745	97.0	251	262.5	1.43	0.60	276.4	1.068
1952	70,882	54,744	83.1	258	274.7	4.65	2.99	278.3	1.060
1953	73,215	66,876	94.3	260	281.6	2.51	3.58	295.0	1.074
1954	74,974	51,755	70.7	261	287.3	2.02	2.27	298.9	1.061
1955	76,326	68,622	91.5	261	292.4	1.78	1.90	300.8	1.047
1956	77,516	67,025	87.8	262	295.9	1.20	1.49	307.5	1.052
1957	81,250	69,978	90.3	265	306.6	3.61	2.41	315.2	1.065
1958	84,496	51,034	62.8	266	317.7	3.62	3.62	342.2	1.116
1959	86,179	53,745	63.6	263	327.7	3.15	3.39	350.2	1.102

[a]Estimated. 1899 estimates from the Census of Manufactures. 1908 estimates from the Directory of Iron and Steel Works.

[b]Predicted from regressions reported in text.

[c]1.067 is the average value of the ratio of B to A in the period 1926-59. It was used to predict output per blast furnace year in 1899.

Source: American Iron and Steel Institute, Annual Statistical Report (New York: AISI, all issues from 1912 to 1960).

141

1. 070 if $CU_t = 1.00$,

1. 082 if $CU_t = 0.789$,

1. 059 if $CU_t = 0.500$,

0. 977 if $CU_t = 0.165$, and

0. 913 if $CU_t = 0.010$.

The ratio is at a maximum for any given value of $\bar{g}_2$ when the rate of capacity utilization is 78. 9 percent.

The fact that output per blast furnace day was found to rise slightly as capacity utilization fell from 100 percent to 79 percent and to fall gradually as capacity utilization fell below 79 percent suggests that managers typically do not take advantage of low capacity periods to operate their most efficient furnaces. Undoubtedly, the major reasons for this finding are connected with uncertainty and the efficiency of operations in other stages of production in vertically integrated plants. If capacity utilization falls very rapidly, uncertainty is introduced about the optimal production plan. Some furnaces may be taken out of blast too soon and may have to be restarted after the passage of a small period of time. Some companies operating multiple plants may decide to shut down plants with efficient blast furnaces because there are inefficiencies in other stages of production.

NOTES

1. J. H. Strassburger, D. C. Brown, R. L. Stephanson, and T. E. Dancy, Blast Furnace—Theory and Practice (New York: Gordon and Breach Science Publishers, 1969), p. 86. Both pellet and sinter particles are distributed approximately normally about a mean particle diameter of one-half inch. The pellets range from one-half to one inch in diameter and more than 50 percent of pellet particles are between one-half and five-eighths inches in diameter. The sinter range is from one-eighth to four inches, with more than 50 percent found in the three-eighths to one inch range.

2. N. D. MacDonald, "The Effect of Screened Sinter on Furnace Capacity, " Proceedings: Blast Furnace, Coke Oven and Raw Materials Committee 20 (1961): 2-15.

3. W. E. Marshall, "Taconite Pellets with Blast Furnace, " Journal of Metals 13, no. 4 (April 1961): 308-13.

4. J. H. Strassburger, et al, op. cit., chapters 2 and 3.

5. Calculated from Directory of Iron and Steel Works (New York: American Iron and Steel Institute, 1967).

6. The values can be found in Table 6.1. The iron content was adjusted upward to allow for the calcined flux in the sinter.

7. "Hi-Pressure Blast Furnace Made, "Journal of Metals 11, no. 12 (December 1959): 787. Remarks attributed to Walter C. Rueckel, division vice president and general manager of Koppers.

8. G. Meinhausen, "Iron and Steel Works—Maximum Capacity, State of Planning, and Chances of Development, " Stahl und Eisen 90, (19 February 1970): 153-61. This was not a U.S. blast furnace.

9. See the last section of Chapter 4. The average injection rate for furnaces actually employing this technique was, of course, much higher. It may be recalled that a 1 percent injection rate is equivalent to 26 pounds of injected fuel per ton of iron produced, or 42 pounds of coke on an equivalent energy basis.

10. W. H. Collison, "Natural Gas Injection at Great Lakes Steel Blast Furnaces, " Iron and Steel Engineer 39, no. 4 (April 1962): 73-81.

11. T. L. Joseph, "The Potential and Limitations of High Blast Temperatures, " Blast Furnace and Steel Plant 49, nos. 3 and 4, (March, April 1961): 239-46, 324-28.

12. "Oil-Coal Injection Saves $1000 a Day on Blast Furnace Coke, " Steel 150, no. 15 (April 1962): 66-69.

13. J. H. Strassburger et al., op. cit., pp. 767-69.

14. Joseph, op. cit.; Collison, op. cit.

15. Meinhausen, op. cit.

16. Blast furnace dimensions and utilization statistics were obtained from the Directory of Iron and Steelworks in America (New York: American Iron and Steel Institute, 1908).

17. See the last section of this chapter for the method employed to estimate output per blast furnace day in the 1899-1925 period. Since 1926, the American Iron and Steel Institute has published this figure in their Annual Statistical Report.

18. These values were estimated from figures on (1) the average number of furnaces per plant, (2) the proportion of merchant furnaces in the total (merchant furnaces sell pig iron to other firms), (3) the durability of furnace linings, (4) the number of blast furnace plants per firm, and (5) the rate of capacity utilization. See Myles C. Boylan, "The Economics of Changes in the Scale of Production in the U.S. Iron and Steel Industry: 1900 to 1970" (Case Western Reserve University, Ph.D. dissertation, 1973).

The labor requirement in a blast furnace plant depends primarily on the labor requirement of each furnace. There is little reduction in hourly employees per blast furnace in multiple furnace plants according to John Fischley of Republic Steel. There are probably small savings in maintenance employees per blast furnace. According to Joseph Johnson in his 1917 publication, the labor input per furnace could be reduced somewhat in multiple furnace plants, but he gave no figures on the magnitude of the saving.[1]

The average driving rate of all furnaces in the plant may have affected the level of employment of hourly workers, particularly in the early decades of this century. If the plant manager decided to operate most of the blast furnaces at low driving rates rather than a smaller number at normal driving rates during periods of low demand, it may have been possible to conserve employment in the casthouse if the intervals between taps were lengthened sufficiently. Small savings in employment in other operations were possible as well. Prior to the widespread diffusion of mechanical charging and casting, considerable labor could be saved in the charging operations at low driving rates. It is unlikely, however, especially after the diffusion of mechanical charging was nearly complete (by 1920), that output per man-hour was reduced on furnaces driven slowly; it is more likely that it rose.

Thus, output per man-hour in the blast furnace industry depends primarily on the size and technical characteristics (for example, blast pressure, mechanization, and automation) of each furnace and the quality of the iron-bearing materials consumed by all furnaces. The remainder of this chapter will be devoted to estimating total man-hours used in the industry and the output per man-hour. Subsequently, increases in output per man-hour will be contrasted with gains in output per blast furnace day (that is, increases in average blast furnace size and the quality of materials) and changes in the utilization of furnace capacity to isolate the contributions stemming from these and other sources.

144

EMPLOYMENT AND PRODUCTIVITY

The Census of Manufactures is the only consistent source of data on employment in the blast furnace industry. In 1947, 1958, and 1963, the census reported total man-hours directly. In all other census years, total man-hours had to be estimated by dividing total wages (reported by the census) by average hourly earnings in blast furnaces (not reported by the census). A second method of estimation would be to multiply total number of hourly employees (reported by the Census) by the average weekly hours worked times 52 weeks. Prior to the 1930s, however, there is not a good time series on average weekly hours worked; and during the last four decades, average weekly hours are reported only for the broader basic steel industry.

Average hourly earnings in basic steel (blast furnaces, steelworks, and rolling mills) are reported in Table 8.1 for the 1899-1963 period. Average hourly earnings in blast furnaces are reported for the years in which they were published: 1899-1935, 1947, 1958, and 1963. Although there may be significant inaccuracies in each series during the first two decades, it may be noted that wages in the blast furnace sector rose relative to wages in the larger, basic steel sector from 1899 to the 1930s. Blast furnace wages were approximately 70 percent of basic steel wages during 1899-1914, 80 percent of basic steel wages during 1919-29, and 90 percent of basic steel wages during 1931-35. Hence, it was assumed that they were 90 percent of basic steel wages in 1935-39. In 1947, wages in the blast furnace sector were slightly below (97 percent) and in 1958 and 1963 slightly above (104 to 106 percent) basic steel wages. Thus, it was assumed that average wages were equivalent in these sectors in the 1949-54 period.

In Table 8.2, the first two columns report data on the average yearly wages per worker and the average number of hourly employees. Column 3 reports the estimates of total man-hours in each census year. As a crude measure of the decline in the work week in the blast furnace sector, it can be noted that total man-hours fell by 50 percent more than the average number of hourly employees from 1899 to 1935-39. Column 4 reports pig iron production in each census year, and column 5 provides estimates of output per man-hour that are illustrated in Figure 8.1.

PRODUCTIVITY AND TECHNOLOGICAL CONSIDERATIONS

It can be noted from Table 8.2 and Figure 8.1 that the most rapid increases in output per man-hour occurred from 1899 to 1909, 1921-29, 1933-39, and 1954-63. The rise during the first decade can be attributed to the large number of new furnaces that were constructed (over

TABLE 8.1

Average Hourly Earnings in Blast Furnaces and Basic Steel: 1899-1963

| | Average Hourly Earnings | | |
	Basic Steel (1)	Blast Furnace (2)	Ratio (2) ÷ (1) (3)
1899	$0.217	$0.158	0.73
1904	0.239	0.174	0.73
1909	0.244	0.179	0.73
1914	0.298	0.206	0.69
1919	0.653	0.512	0.78
1921	0.507	0.409	0.81
1923	0.579	0.483	0.83
1925	0.617	0.517	0.84
1927	0.625	0.522	0.84
1929	0.635	0.528	0.83
1931	0.617	0.551	0.89
1933	0.509	0.467	0.92
1935	0.651	0.577	0.89
1937	0.818	n.a.	(0.90)[*]
1939	0.842	n.a.	(0.90)
1947	1.45	1.41	0.97
1949	1.66	n.a.	(1.0)
1950	1.70	n.a.	(1.0)
1951	1.90	n.a.	(1.0)
1952	2.00	n.a.	(1.0)
1953	2.18	n.a.	(1.0)
1954	2.22	n.a.	(1.0)
1958	2.88	2.99	1.04
1963	3.31	3.51	1.06

[*]Data not available. The ratio was assumed to be 0.9 in 1937-39 and 1.0 in 1949-54.

Sources: Column 1, 1899-1919 from Paul Douglas, Real Wages in the U.S. 1890-1926 (New York: A.M. Kelley, 1966); 1921-33 from the National Industrial Conference Board (figures were adjusted downward slightly to compensate for the NICB's sample, which was biased toward high wage regions); 1935-63 from the American Iron and Steel Institute Annual Statistical Report (New York: AISI, indicated issues during 1935-63). Column 2, 1899-1931, U.S. Bureau of Labor Statistics index of weighted average of all wage-earning positions in the blast furnace industry, reported in the 1932 Statistical Abstract of the U.S. (Washington, D.C.: U.S. Bureau of Foreign and Domestic Commerce, 1932). Values were linked to 1929-35 actual wages; 1929 wage from "BLS Bulletin No. 513" (1929); 1931 wage from "BLS Bulletin No. 567" (1931); 1933-35 wages reported in C.R. Daugherty, M.G. DeChazeau, and S.S. Stratton, Economics of the Iron and Steel Industry (New York: McGraw-Hill, 1937); 1947, 1958, and 1963 estimated directly from data on total man-hours and total wages in blast furnaces, reported by the Census of Manufactures (Washington, D.C., U.S. Department of Commerce, Bureau of the Census, 1947, 1958, 1963).

TABLE 8.2

Total Man-Hours, Output, and Output per Man-Hour

Year	Average Yearly Wages per Wage Earner (1)	Average Number of Wage Earners (2)	Total Number of Man-Hours (thousands) (3)	Pig Iron Output (Gross Tons $\times 10^6$) (4)	Output per Man-Hour (5)
1899	$ 470	39,358	117,090	13.6	0.12
1904	539	35,178	108,981	16.2	0.15
1909	640	38,429	137,384	25.4	0.19
1914	776	29,356	110,584	23.0	0.21
1919	1,776	43,296	150,294	30.5	0.20
1921	1,570	18,698	71,782	16.4	0.23
1923	1,605	36,712	121,994	39.7	0.33
1925	1,552	29,188	87,622	36.1	0.41
1927	1,583	27,958	84,797	35.9	0.42
1929	1,681	24,960	79,473	41.8	0.53
1931	1,419	13,572	34,948	18.0	0.51
1933	956	12,098	24,765	13.0	0.52
1935	1,246	15,178	32,769	20.8	0.63
1937	1,647	23,075	51,642	36.1	0.70
1939	1,449	19,537	37,355	31.1	0.83
1947	2,843	32,697	66.065	52.1	0.79
1949	3,079	31,911	59,163	47.7	0.81
1950	3,254	35,357	67,677	57.7	0.85
1951	3,738	38,062	74,882	62.7	0.84
1952	3,605	38,201	68,857	54.7	0.80
1953	4,358	38,809	77,582	66.9	0.86
1954	4,194	31,108	58,769	51.8	0.88
1958	5,838	20,713	40,391	51.0	1.26
1963	7,339	19,655	41,060	64.1	1.56

Sources: Columns 1 and 2, 1899-1954 from the Census of Manufactures (Washington, D.C.: Bureau of the Census, all issues from 1899 to 1963); 1958 and 1963 wages estimated directly from census data on total wages and total man-hours in the blast furnace sector. Then average weekly hours in basic steel, reported by the U.S. Bureau of Labor Statistics, was used to derive yearly wages per wage earner; and this was divided into total wages to derive the average number of wage earners.

Column 3, 1899-1935 estimated by dividing total wages [(1) × (2)] by average hourly earnings in the blast furnace sector; 1937 and 1939 estimated by dividing total wages by average hourly earnings in basic steel times 0.9 (see Table 8.1); 1949-54 total wages were divided by average hourly earnings in basic steel; 1947, 1958, and 1963 reported directly by the Census of Manufactures (Washington, D.C.: Bureau of the Census, 1947, 1958, 1963).

Column 4, pig iron includes output of charcoal fueled furnaces in the 1899-1929 period. Figures reported in the American Iron and Steel Institute's Annual Statistical Report (New York: AISI, selected issues from 1911 to 1963).

FIGURE 8.1

Output per Man-Hour
(Gross Tons per Man-Hour)

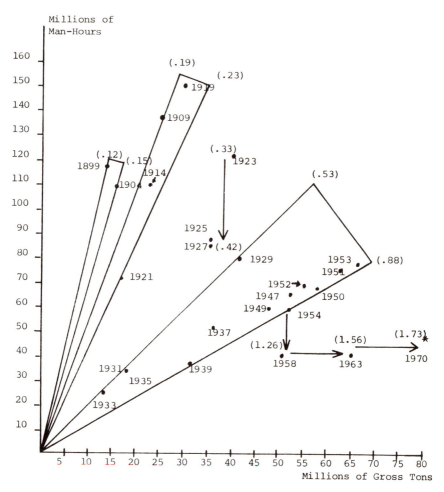

Source:Table 8.1.

100). These new furnaces were much larger than the average size of existing furnaces; and they were more mechanized, for example, they were designed for mechanical charging and casting rather than hand charging and casting. Thus, even though the labor requirements of larger furnaces may have been greater than that of smaller furnaces with <u>similar</u> technical characteristics, the newer furnaces contributed greatly to lower labor requirements in the industry because they were far more mechanized than smaller and older units.

The major increase in output per man-hour during the 1920s can be attributed to the scrapping of many inefficient furnaces, the progressive enlargement and improvement of other furnaces, and improved operating "knowhow."[2] The improvement made during the 1930s was due to similar developments but also can be attributed to the low capacity utilization rates that prevailed during most of that decade. For example, pig iron production fell drastically from 36 million tons in 1937 to 18.5 millions in 1938 and rose again to 31 million tons in 1939. As demand increased during late 1938 and 1939, only the most efficient furnaces were started up, with the result that output per man-hour rose from 0.70 in 1937 to 0.83 in 1939.

Very little improvement in average labor productivity was observed from 1939 to 1954, primarily because most furnaces were in blast during those years, limiting the extent to which they could be rebuilt; and very few furnaces were scrapped. There was also a deterioration in the natural quality of iron-bearing materials during this period, resulting in construction of additional sintering capacity in blast furnace plants and requiring additional employees to process the materials in sinter plants.

From 1954 to the present there was a major increase in output per man-hour due to significant improvements made in the quality of the iron-bearing materials, increases in average blast-furnace size, and (in recent years) greater automation in furnace operations. A major increase in sintering capacity during this period offset to some extent the gains in output per man-hour in other operations.

It is interesting to note that William Hogan's estimates of increases in output per man-hour from 1920 to 1946 in a major blast furnace plant (which he called plant A)[*] closely paralleled the increases estimated in this study.[3] Because plant A operated furnaces that were (on the average) considerably larger in size than furnaces operated in the industry, average labor productivity there was 33 to 64 percent higher.

If the labor requirement per blast furnace were constant, increases in output per blast furnace day would contribute directly to increases in output per man-hour. Since the labor requirement per blast furnace is likely to rise (although slowly) as the furnace is increased in size

---

[*]Plant A is believed widely to be the Edgar Thomson plant of the United States Steel Corporation, located in Pittsburgh.

or the materials throughput on a furnace of constant size is increased (due to improvements in the quality of the iron-bearing materials), increases in output per blast furnace day contributed to somewhat smaller increases in output per man-hour. From 1899 to 1929, output per blast furnace day increased by 70 percent of the gain in output per man-hour, from 1929 to 1954 by 64 percent of the gain, and from 1954 to 1963 by 69 percent. Thus, increased mechanization and automation of tasks previously performed by hourly workers contributed to more than 30 percent of the gain in output per man-hour. Part of this contribution probably stemmed from the substitution of salaried employees for hourly labor.

NOTES

1. Johnson, Blast Furnace Construction in America (New York: McGraw-Hill, 1917), p. 8.
2. William T. Hogan, S.J., Productivity in the Blast Furnace and Open Hearth Segments of the Steel Industry: 1920-1946 (New York: Fordham University Press, 1950), pp. 53-70.
3. Ibid.

9

SOME EVIDENCE ON ECONOMIES OF SCALE IN THE IRON INDUSTRY

SOURCES OF INCREASE IN THE AVERAGE SIZE OF BLAST FURNACES

The increases in the average size of blast furnaces (reviewed in Chapter 8) were the result of three activities: the construction of new furnaces, the scrapping of old furnaces, and the rebuilding of existing furnaces to larger dimensions.

A furnace is defined by the industry to be newly constructed when the entire furnace and its ancillary equipment are new. The rebuilding activity encompasses a range from minor increases in the internal dimensions of the furnace and no change in the ancillary equipment (usually accomplished then the furnace is relined) to major increases in the size of the furnace and concomitant increases in the capacity of the ancillary equipment. For example, a major increase in furnace dimensions (an increase of five feet in the hearth and bosh diameters) may require (1) dismantling the original furnace down to its foundations, strengthening the foundations, and constructing a new shell and top; (2) replacing the original buckets of the skip hoist with larger units; (3) increasing the stove capacity by any combination of increased maximum temperature capability in the stoves' combustion chambers, enlarged stoves,* and the addition of a new stove; (4) increasing the capacity of the turboblowers by replacement; and (5) replacing the iron removal ladles with larger units. Minor rebuilds have been more numerous than major rebuilds because ". . . most furnaces are enlarged when they are being relined."[1]

---

*The stoves may be enlarged externally by building larger shells or they may be enlarged internally (in the sense of increasing the square feet of heating area) by decreasing the thickness of the firebrick and decreasing the width of the interstices in the checkerwork of bricks. The latter improvements could be made as improvements were made in the quality of the firebrick and gas cleaning methods.

The industry designates permanently shut down furnaces as "abandoned." If an abandoned furnace is immediately torn down, it is instead designated as "dismantled." In this study, both abandoned and dismantled furnaces are considered to be scrapped.

New and Scrapped Furnaces

Figure 9.1 indicates the extent of new furnace construction since 1900. It is evident that the majority of new furnaces constructed in the 1900-70 period were built during the first two decades. Of the 258 new furnaces constructed, 121 were built in 1900-09 and 56 were built in 1910-19. It is also evident that the variation in the hearth diameters of furnaces newly built within a few years of each other was very large in the early 1900s, but decreased both relatively and absolutely through time, particularly if furnaces constructed in the southern and western regions of the country and furnaces producing ferroalloys and other special iron products are eliminated from the comparison.[2] Finally, it should be noted that there were two major increases in the trend line representing the size of the largest new furnaces. One occurred in the late 1920s and early 1930s, and the other occurred during the late 1960s.

Figure 9.2 illustrates the frequency and size of furnaces that were scrapped in the period 1908-70.[*] The greatest concentration of scrapping occurred from 1920 to 1939, when 210 furnaces were scrapped (a total of 350 furnaces were scrapped in the entire 1908-70 period).[†] In contrast, only 22 furnaces were (permanently) scrapped

---

[*] Bosh diameter was used as the indicator of size because it was the only dimension continuously available for this period. There was too much possible variation in the bosh diameter-hearth diameter differential to attempt an estimate of each furnace hearth diameter in the 1908-30 period—as was done in Figure 9.1.

[†] The scatter diagram of scrapped furnaces in Figure 9.2 is based on the firms' decisions to announce the scrapping of these furnaces. In 1925, the American Iron and Steel Institute (AISI) decided not to count the capacities of those furnaces that had been idle for a number of years (and were unlikely to be used in the future) in their estimate of industry capacity. As a result, the AISI reported 31 more furnaces scrapped between 1925 and 1929—and 6 more furnaces scrapped from 1925 to 1939 than were formally abandoned or scrapped during those periods. Table 7.6 was based on the AISI figures for building and scrapping activity (since that chapter dealt with average furnace capacity, based on the AISI estimate of capacity). Thus, Figure 9.2 should indicate 31 less scrapped blast furnaces in the 1920s and 25

FIGURE 9.1

Scatter Diagram of the Hearth Diameters
of New Blast Furnaces

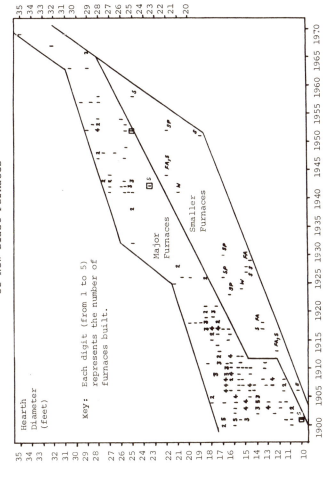

Note: In the size range of smaller furnaces: S=South, FA=Ferroalloys, W=West, and SP= producers of specialty irons in the North.

Source: Directory of Iron and Steelworks in America (New York: American Iron and Steel Institute, all issues since 1908).

153

FIGURE 9.2

Scatter Diagram of the Bosh Diameters of
Scrapped Blast Furnaces

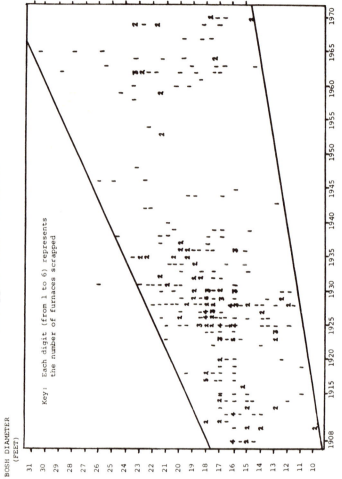

Source: Directory of Iron and Steelworks in America (New York: American Iron and
Steel Institute, all issues since 1908.)

154

in the 1940-59 period.[*] There was enormous diversity in the size of furnaces scrapped within a few years of each other. The relative variance in their bosh diameters increased slightly, and the absolute variance doubled from 1908 to 1970.

There was a reasonably distinct boundary between the sizes of new and scrapped furnaces. In any given year except 1931, all new major furnaces (indicated as such in Figure 9.1) were larger than all of the furnaces scrapped. The smaller new furnaces (plotted in Figure 9.1) were often equalled or exceeded in size by some of the scrapped furnaces.

If the capacities of new and scrapped blast furnaces are averaged for each year in the shorter 1912-59 period in which furnace capacities were reported (to remove the intrayear variance in capacity), most of the influence exerted by differences in furnace location and managerial preference about size should be removed. (In some years there was insufficient scrapping and building activity to remove entirely the influence of these factors.) Most of the remaining interyear variance should reflect the gradual tendency of new furnaces and scrapped furnaces to grow in size and capacity as increasing "knowhow" and technological advances remove obstacles to increasing the dimensions of the largest blast furnaces. To indicate the reduction in variance and to estimate the tendency for new and scrapped furnaces to grow in capacity over time, the average yearly capacities of new and scrapped furnaces were plotted in Figure 9.3; and these values were fitted to "time" by estimating the following weighted regression equations (the weights are based on the relative number of furnaces built or scrapped in each year):

New furnaces:

$$\hat{CAP} = 112,300 + 7,480 \, (Date - 1912), \quad R^2 = 0.85 \qquad (9.1)$$

Scrapped furnaces:

$$\hat{CAP} = 23,600 + 3,900 \, (Date - 1912), \quad R^2 = 0.67 \qquad (9.2)$$

where $\hat{CAP}$ represents the average capacity per year in gross tons.

The fit between average yearly capacity and time was high for new

---

more scrapped furnaces during the 1930s than Table 7.6. In fact, it reports 40 less during the 1920s and 20 more during the 1930s, due to lack of information on 14 of the blast furnaces scrapped during these two decades.

[*]A number of other furnaces were reported as abandoned by their owners, but were subsequently operated by either the original or a new owner as the result of a high demand for iron and steel products during most of the years of this period.

FIGURE 9.3

Average Yearly Capacities of New and Scrapped
Blast Furnaces: 1912-59

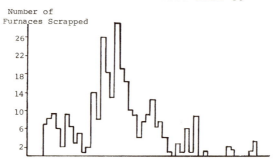

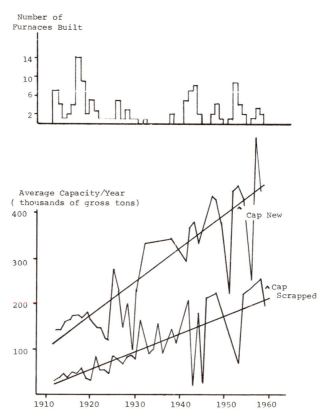

Source: American Iron and Steel Institute, Annual
Statistical Report (New York: AISI, all issues from 1911-59)

furnaces—85 percent of the interyear variance in capacity was explained by time—and the explained variance would have been higher if some of the unrepresentative observations had been deleted. The fit for scrapped furnaces was moderate—67 percent of the variance in capacity was explained by time, although it too would have been higher if some of the smaller, specialty furnaces had been deleted from the regression.

Because the slope parameter for new furnaces is almost twice that for scrapped, the absolute difference in capacities of new and scrapped furnaces has tended to grow over time. Even so, the relative difference has tended to decline. In 1912, the capacity of one new furnace replaced, on the average, the capacity of almost five scrapped furnaces; but in 1959 one new furnace replaced slightly more than two scrapped furnaces.

Rebuilt Furnaces

From 1900 to 1970, about 380 coke-burning blast furnaces were scrapped. In 1900, there were approximately 350 blast furnaces. Thus, the capital stock was turned over slightly more than once. Of the 260 furnaces constructed since 1900, 180 or 70 percent were built prior to 1920, and only one was as large as 20 feet in hearth diameter. Yet, the weighted average hearth diameter of utilized furnaces was estimated in Chapter 7 to rise from 15-1/2 feet in 1919, to 20-3/4 feet in 1929, and (more slowly) to 26-1/3 feet in 1970.[3] It is thus apparent that rebuilding was an important source of increases in blast furnace size.

The relative significance of rebuilding as a third source of additional capacity can be approximated by adjusting the total yearly capacity of U.S. furnaces reported by the American Iron and Steel Institute in the 1914-54 period for the individual capacities of new and scrapped furnaces and attributing the residual to rebuilding.[4] This technique is likely to overestimate this source of additional capacity because it ignores improved operating "knowhow" and other factors influencing capacity.*

From 1914 to 1954, industry capacity rose by 71.3 percent from 49 million to 84 million net tons. Four-fifths of this increase (58.5 percent) was due to rebuilding and the remainder was due to the excess capacity of 123 new blast furnaces over 278 scrapped furnaces. An average of 6 percent additional capacity was added by rebuilding in each five-year period, varying from a low of 1.5 percent in 1945-49 to a high of 13.2 percent in 1925-29.

---

*On the other hand, some rebuilding is necessary simply to maintain the existing capacities of furnaces due to physical wearing.

The effect of rebuilding on average furnace capacity is reported in Table 9.1. Beginning with an average capacity of 120,000 net tons per year in 1914, average capacity reached 322,000 net tons by 1954. More than one-half of this increase resulted from rebuilding.

ECONOMIC, TECHNICAL, AND MANAGERIAL FACTORS INFLUENCING BLAST FURNACE SIZE

Building New Furnaces or Rebuilding Existing Ones

The demand for pig iron is primarily a derived demand. Managers of integrated plants typically determine the need for additional pig iron capacity on the basis of expected future sales of the plant's finished steel products.[5] When additional capacity is needed, the decision on how to acquire it and how much to expand is influenced by (1) the expected growth in (regional) demand for the finished products of the plant, (2) plant size, and (3) cost considerations.

The objective of managers is to minimize the total production cost per year for an expected volume of production. The faster the future growth in demand is expected to occur, the greater are the additions to capacity that will be planned.

Additional capacity may be provided by either rebuilding existing furnaces or building new furnaces. The costs associated with rebuilding are the losses resulting from low rates of operation while a furnace is taken out of blast and the cost of the rebuilding itself, less the present value of future savings in operating cost after the furnace is rebuilt. The costs of building a new furnace are the amount of investment expenditures, less the present value of future savings in operating costs.[*]

The available evidence, although not very good,[†] indicates that the unit capacity cost for new furnaces has not been substantially different from the unit capacity cost of the extra capacity derived from

------

[*]In both the rebuilding and new furnace cases, greater excess capacity will be present after construction is completed (either in the plant itself or in some other plant owned by the parent firm), assuming that the total demand for the firm's products increases relatively smoothly rather than in large jumps. Since the new or rebuilt furnaces will be fully utilized, they not only fill the increases demand but also partially displace operations on less efficient furnaces —giving rise to some savings in operating costs.

[†]The data are suspect in some cases and there were an inadequate number of observations. Also, the unit capacity cost of rebuilding is likely to exhibit wide variance, depending on the extent of the rebuilding and the condition of the furnace being rebuilt.

TABLE 9.1

Increases in Average Furnace Capacity over Five-Year Intervals due to
New Building, Scrapping, and Rebuilding of Blast Furnaces
(all capacity figures in thousands of net tons)

Year	Total Capacity on Dec. 31 (1)	Total Capacity on Dec. 31 If No Rebuilding During Previous 5 Years (2)	Number of Blast Furnaces on Dec. 31 (3)	Actual Average Furnace Capacity Per Year (1) ÷ (3) (4)	Average Furnace Capacity If No Rebuilding During Previous 5 Years (2) ÷ (3) (5)	Five-Year Increases in Average Furnace Capacity Attributable to Scrapping and New Construction (6)[a]	Rebuilding (7)[b]
1914	49,007	—	409	119.8	—	—	—
1919	55,583	53,956	421	132.0	128.2	8.4	3.8
1924	59,244	55,837	401	147.7	139.0	7.0	8.7
1929	57,669	50,161	277	208.2	175.4	27.7	32.8
1934	56,948	54,314	251	226.9	211.7	3.5	15.2
1939	55,628	53,253	222	250.6	232.0	5.1	18.6
1944	67,250	62,272	241	279.0	264.7	14.1	14.3[c]
1949	71,458	70,452	246	290.5	288.2	9.2	2.3[d]
1954	83,971	78,817	261	321.7	307.5	17.0	14.2
Totals						92	109.9

[a] Derived by subtracting column 5 from the preceding row value in column 4.

[b] Derived by subtracting column 5 from column 4 (in the same row).

[c] Likely an overestimate because furnaces were driven very hard during the war years.

[d] Likely an underestimate because furnaces returned to normal driving rates after the war years and because the quality of the iron-bearing materials declined in this period.

Source: American Iron and Steel Institute, Annual Statistical Report (New York: AISI, various issues).

159

rebuilding an existing unit. These costs are compared in Table 9.2 and Figure 9.4.

In contrasting rebuilding with new building, rebuilding will be selected when losses resulting from low rates of operation are smaller than the amount by which the capital cost of the new furnace exceeds the capital cost of rebuilding (net of the present value of likely savings in future operating costs in each case). The latter difference in capital costs basically depends on the higher (excess) capacity of the new furnace, according to the data presented in Table 9.2.

One important conclusion is that, as the desired increase in capacity grows relative to the size of the plant, the losses resulting from low rates of operation during rebuilding are likely to grow, and the difference between the capital outlays for a new furnace and rebuilding is likely to diminish, increasing the likelihood that a new furnace would be constructed.

Another notable conclusion is that in situations where a new furnace is selected in preference to rebuilding, the new furnace probably will be larger in large plants. In a single furnace plant, either source of new capacity will be costly. Rebuilding will result in a total loss of capacity for a number of months; but a new furnace will at least double capacity, possibly creating high excess capacity. As larger plant sizes are considered, both of these penalties are diminished in magnitude. In a plant with three or more furnaces, the desired increment to capacity can be achieved by a number of rebuilding patterns (for example, minor rebuilds of each furnace at different points in time or a major rebuild of one furnace) that may permit the plant to maintain normal operations. If a new furnace is added, excess capacity will be minimal and it is more likely, ceteris paribus, that the plant has an old and small furnace approaching obsolescence — leading management to decide to build the largest possible new furnace if they choose that source of expansion.

The figures in Table 9.1 in conjunction with Figure 9.5 support the conclusions of this analysis. The growth in the importance of rebuilding as against new construction as a source of new capacity jumped during the 1920s when the derived demand for pig iron exhibited no growth. It increased further in relative importance during the 1930s when demand was depressed and dropped somewhat in relative significance during the 1940s as demand rapidly rose and stabilized. It did not drop in significance during the 1950s despite further increases in demand. But average plant size grew significantly during 1914-39, increasing the flexibility of managers to choose between rebuilding and new construction in later years. Over the broader 1908-66 period, the number of single furnace plants fell from 132 to 15, the number of two furnace plants dropped from 54 to 19, and the number of plants with three or more furnaces remained constant in the mid-1930s.

TABLE 9.2

Investment Cost per Gross Ton of Capacity: New and Rebuilt Furnaces

	New			Rebuilt			
Period or Year	Number of Observations	Average Annual Capacity per Furnace	Average Cost per Gross Ton (dollars)	Period or Year	Number of Observations	Average Addition to Capacity	Average Cost per Gross Ton (dollars)
1903–11	7	130,000	3.45	—	—	—	—
1916–21	6	200,000	3.60	1913	1	90,000	3.35
1926	1	250,000	9.20	1924–30	11	95,000	11.25
1939–44	7	360,000	12.40	1934–40	9	110,000	10.40
1952–53	2	455,000	21.80	1950–60	7	40,000	14.05
1964	1	635,000	33.15	1962–68	9	80,000	28.40

Note: Capacity figures were normalized for differences in materials quality.
Sources: Data supplied by Armco Steel, Inland Steel, and Republic Steel. Figures on capital costs of United States Steel were taken from William T. Hogan, Productivity in the Blast Furnace and Open-Hearth Segments of the Steel Industry: 1920–1946 (New York: Fordham University Press, 1950), pp. 45–50.

FIGURE 9.4

Investment Cost per Gross Ton of Capacity
for New and Rebuilt Furnaces

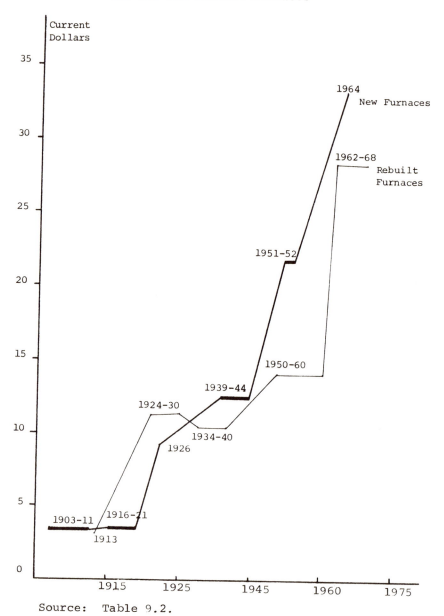

Source: Table 9.2.

FIGURE 9.5

Pig Iron Output and Industry Capacity

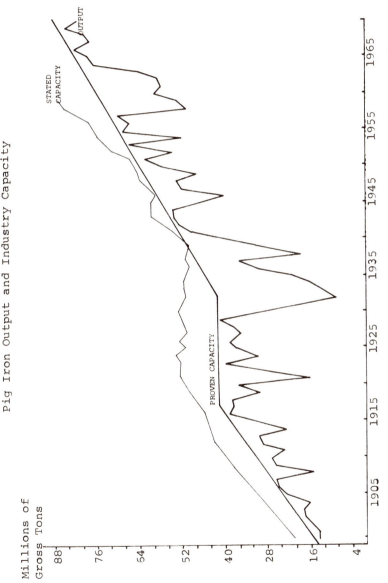

Source: American Iron and Steel Institute, Annual Statistical Report (New York: AISI, various issues).

163

The Choice of New Furnace Size

<u>Technical Considerations</u>

In the early part of the century, smelting iron was largely an art. Blast furnace operators learned through years of experience how to determine what was happening inside the furnace and, thus, how to alter the proportions of iron-bearing materials, coke, and flux and how to adjust the wind rate and blast temperature to produce pig iron of acceptable quality.

Because a large investment in time was necessary to achieve the requisite expertise, operators may have been reluctant to supervise furnaces outside their range of experience. Occasionally a firm would build a furnace larger than all others. If operations on that furnace were successful, that is, if the quantity of materials and fuel consumed per unit of output did not rise and the product was acceptable in quality, then other firms would follow the innovator to the extent that industry growth and cost considerations permitted. In short, innovations in size appear to have been determined largely by engineers, while the diffusion of the innovation depended on economic considerations.

The major jump in the upper boundary of new furnace size (see Figure 9.1) in the late 1920s was due to this process, according to Hogan.

> From 1875 to 1920, the growth of the blast furnace was accomplished by gradual evolution. This terminated in 1918 with the construction at the South Works, Chicago, of a furnace with a hearth 20 feet, 9 inches, in diameter and a stack 92 feet high. The average large furnace of that period, however, had an 18-foot hearth and a stack about 90 feet high. It was the next 25 years (1920-1945) that ushered in the era of the big furnace as we know it today (circa 1950).
>
> The results obtained (from the 20-foot, 9 inch, hearth furnace) at the South Works were so satisfactory that the unprecedented step was taken in 1924 of building a furnace 103 feet high, with a 25-foot hearth, at Youngstown, Ohio. This was a revolution in furnace design. It operated with such eminent success, however, producing 1092 net tons of iron per day with a coke consumption of less than 1600 lbs. per ton of iron, that it resolved all doubts about the efficiency of large furnaces.
>
> Thus, after 1929, the large furnace was established not only as a feasible but also an economical unit. However, a few years expired before there were a number of them in operation. The reason for this delay was twofold. First, the demand for pig iron dropped off drastically in

the early 1930's, due to general economic conditions. . . .
Therefore, little thought was given to building larger furnaces during these years. Second, . . . there was very
little need for relining at this time, because most furnaces
were inactive and did not wear out. Therefore, very few
furnaces "grew" in size during the depression.[6]

The growth in the size of the largest furnaces was relatively slow
again during the next three decades. Until the early 1960s, there was
a prevailing belief that blast furnaces with hearth diameters much
larger than 28 feet would not perform well.[7] Apparently, it was founded
on operating experience with natural domestic ore and on the related
Rice ring concept (which is based on the assumption that the air blast
cannot penetrate beyond six feet into the upper hearth area of the
furnace). The development of high quality agglomerates in the late
1950s paved the way for rendering this theory obsolete.

The major jump in the late 1960s occurred as the leadership in
the design of large furnaces passed from the United States to Japan
and the Soviet Union. In 1963, Armco Steel constructed a furnace
with a hearth diameter of $30\frac{1}{2}$ feet, which was claimed to be the
world's largest furnace. But by the time Bethlehem Steel's 35-foot
hearth furnace was in operation in 1970, the Japanese were operating
successfully furnaces with hearth diameters in excess of 40 feet.
Data explaining this lagging behavior of U.S. firms will be presented
later in this chapter.

Economic Factors

The underlying assumption in this section is that managers wish
to minimize the long run average cost of producing a given quantity of
iron. Unfortunately, data pertaining to the possible cost advantage of
large blast furnaces are not available on an industrywide basis over
the entire period under study. For this reason, the operating and
capital costs during the 1960s in the Pittsburgh to Chicago region will
be pieced together. In this section, it will be shown that managers
preferred large blast furnaces, and some justifications for the variance
in the size of new furnaces (constructed within a few years of each
other) will be offered.

If one accepts the premise of Stigler's "survivor principle:" that
firms over the long run adjust their scale of plant to minimize costs,[8]
then the evidence supports the claim that large blast furnaces are more
efficient. In every case where a blast furnace plant expanded its
capacity by constructing a new furnace, the new furnace was greater
than or equal to the size of existing furnaces in the plant. Given this
managerial preference for large furnaces, the question is, why was
there variance in the size of these furnaces ?

In the early decades of this study, several factors appear to have
been important. Managerial concern with cost was focused less sharply

than it is today. Not only were techniques of cost analysis less well developed, but also the industry was dominated by engineers who were most concerned with product quality. This precluded daring gambles with ever larger new furnaces, except by the largest firms that could survive possible negative consequences.

Probably even more important in the earlier years were the incentives to locate in certain specific regions of rapid growth. The first two decades of this century represented the second half of a major expansion of capacity and a shift in location from the east to the eastern midwest. Sixty-five percent of the new furnaces built from 1900 to 1920 were located in either the Pittsburgh, Youngstown and upper Ohio River region (which advantageously is located near deposits of coking coal), or on the Great Lakes, primarily Buffalo, Cleveland, and Chicago (which are advantageously located in respect to the Lake Superior ore deposits). The major growth in capacity in these areas was in sharp contrast with the more limited growth in other northern regions and in the south and west. It may be surmised that the "slow growth" regions grew slowly because of poor location with respect to markets and/or the limited ability of firms to expand the production of underground ore mines serving those regions. As was demonstrated earlier in this section, all but the largest plants in slow growth regions are disinclined to build the largest possible blast furnaces; plants in rapid growth regions are likely to build large furnaces, since that is the fastest way to expand capacity. This consideration alone should explain a large part of the variance in new furnace sizes.

The major factor underlying the variance of growth of different regions is that average total production cost was so sensitive to location. A comparison of published raw materials transportation costs in the high growth regions—Pittsburgh, Youngstown, Buffalo, Cleveland, and Chicago (see Table 9.3 and Figure 9.6)—indicates that, even in these cities, there was a significant variance in the costs of transporting materials per ton of iron produced; and these costs constituted major portions of the value of the pig iron produced. During the 1930s, approximately 40 percent of the value of iron produced represented the transportation costs of raw materials. (This cost proportion dipped to 30 percent in 1950 and fell more gradually to 20 percent by the late 1960s.) Since value added was generally less than 20 percent during the 1900-70 period, these costs were at least as important as capital and labor costs combined.

During the 1930s there was a spread of $1.13 (in 1934) to $1.28 (in 1938) in transportation costs to cities included in Table 9.3 and probably higher transportation costs elsewhere, while the typical ton of iron was valued at $16, according to the U. S. Census of Manufactures. If these data are applicable to earlier periods on a proportionate basis, it is very doubtful that even the construction of very large new furnaces (assuming average total cost falls for larger furnaces) in locations with high transportation costs for raw materials (which are not included in Table 9.3) could have offset their locational

TABLE 9.3

Transportation Costs for Raw Materials Consumed
per Ton of Iron at Selected Locations

		1934	1938	1950	1960	1970
Pittsburgh:	Ore	5.77	5.80	9.59	11.28	12.18
	Coal	0.65	0.28	0.63	(0.65)*	(0.65)*
	Total	6.42	6.08	10.22	11.93	12.83
Youngstown:	Ore	5.13	5.19	8.69	10.17	10.97
	Coal	2.07	1.98	2.90	2.66	2.53
	Total	7.20	7.17	11.59	12.83	13.50
Buffalo:	Ore	3.26	3.50	n.a.	6.78	7.31
	Coal	2.81	2.91	n.a.	5.05	4.45
	Total	6.07	6.41		11.83	11.76
Cleveland:	Ore	3.26	3.50	6.05	6.78	7.31
	Coal	3.06	2.71	5.00	5.03	4.45
	Total	6.32	6.21	11.05	11.81	11.76
Chicago:	Ore	3.25	3.49	6.05	6.78	7.31
	Coal	3.35	3.87	5.98	6.03	5.08
	Total	6.60	7.36	12.03	12.81	12.39

*Estimated.

Sources: 1934, C.R. Daugherty, M.G. DeChazeau, and S.S. Stratton, Economics of the Iron and Steel Industry, vol. 1 (New York: McGraw-Hill, 1937), table 80, p. 378; 1938, Marion Worthing, "Comparative Assembly Costs in the Manufacture of Pig Iron," Pittsburgh Business Review 8, No. 1 (1 January, 1938): table 1, pp. 21-25; 1950, H.H. Chapman, Iron and Steel Industries of the South (Tuscaloosa: University of Alabama Press, 1953), table 61, pp. 193-97 (the burden rate is assumed to be 2.0 and the coke rate to be 0.90); 1960 and 1970, calculated from published freight rates for iron ore and coal, assuming the burden rates to be 1.72 and 1.68 and the coke rates to be 0.76 and 0.63 in 1960 and 1970, respectively.

FIGURE 9.6

Transportation Costs for Raw Materials in
Selected Cities (per Gross Ton of
Pig Iron)

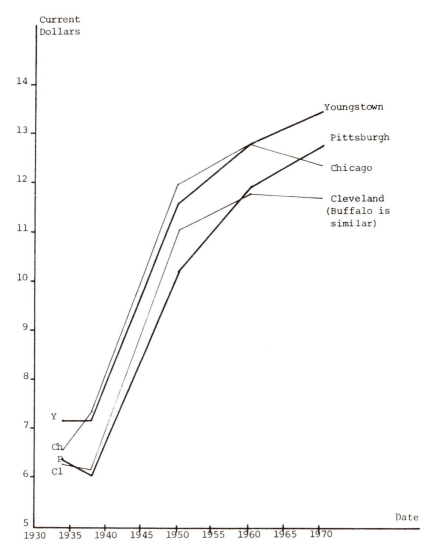

Source: Table 9.3

disadvantage, permitting them to penetrate effectively the markets of the low transportation cost centers (with finished steel products). Hence, the high transportation cost regions were slow growth regions because they were restricted to serving local customers and were subject to competition from more advantageously located plants.

After the early 1920s, the variance in the sizes of new furnaces diminished both absolutely and relatively over time (as noted in Figure 9. 1). The remaining variance can be explained by the fact that some furnaces were built to produce special grades of pig iron for which demand was limited and by the continued slow growth of some regions, particularly the south (see Figure 9. 1 and note the characteristics of furnaces labeled "smaller furnaces").

Scrapped Furnaces

Despite the large variation in sizes of scrapped blast furnaces, there is again ample support for the theories that managers preferred larger furnaces and, under Stigler's premise, that restricting operations to larger furnaces reduced the average total cost of the plant.

In the 1908-66 period, 338 furnaces were scrapped. The scrapping of 226 furnaces was accompanied by the shutting down of 174 plants. (Presumably, small plants failed to survive for reasons noted earlier in this section.) The other 112 furnaces were compared in size with furnaces retained in the same plant on a "per blast furnace retained" basis.* The average plant size before scrapping, in those instances where comparisons were possible, was larger than four furnaces, permitting 383 comparisons to be made. The results are reported in Table 9. 4.

First, note that few decisions to scrap were concomitant with decisions to increase plant capacity, suggesting that most furnaces are scrapped for reasons relating solely to their technical and economic performance. Second, in those cases where intraplant comparisons were possible and plant capacity decreased, retained furnaces were smaller than scrapped furnaces with a frequency of only $1\frac{1}{2}$ percent. Third, in those cases where plant capacity increased despite scrapping, retained furnaces were smaller than scrapped furnaces about 10 percent of the time. But in most of those cases, the difference in size was small; or the smaller furnaces were retained to produce ferroalloys, which are needed in only limited quantities. Overall, in only $3\frac{1}{2}$ percent of the comparisons were scrapped furnaces larger than retained furnaces.

---

*As opposed to a "per blast furnace scrapped" basis. In situations where two (or more) furnaces were retained, one smaller and the other larger than the scrapped furnace, the latter basis would not give a clear comparison.

TABLE 9.4

Intraplant Comparisons Between Retained
and Scrapped Blast Furnaces, 1908-66

Overall Change in Plant Capacity	Total Number of Furnaces Scrapped	Number of Comparisons Between Retained and Scrapped Furnaces Where Retained Furnaces Were[b]:		
		Smaller	Same Size	Larger
Decreased	85	4 $(1\frac{1}{2}\%)$	35 (12%)	251 $(86\frac{1}{2}\%)$
Increased due to re-placement, or rebuild-ing and new construc-tion elsewhere	27	9 $(9\frac{1}{2}\%)$	8 $(8\frac{1}{2}\%)$	76 (82%)
Total	112	13 $(3\frac{1}{2}\%)$	43 (11%)	327 $(85\frac{1}{2}\%)$

[a]From preceding year to succeeding year.
[b]Bosh diameters were used as indicators of size until 1935. There-after hearth diameters were used.
Source: Directory of Iron and Steelworks in America (New York: American Iron and Steel Institute, various issues).

The primary factor causing large fluctuations in the size of scrapped furnaces in the industry was location. Only 23 percent of the 282 furnaces scrapped from 1908 to 1939 were located in the Pittsburgh, Youngstown, and upper Ohio River region, or on the Great Lakes; but 77 percent were located in other areas in the north, the south, and the west. Because of the rising availability of foreign ore and agglom-erates during the 1950s, these proportions changed to 50 percent each in the 1940-70 period, during which 68 furnaces were scrapped. (There was also a small but significant shift in U.S. markets during the latter period to the east and west coasts, the south and the Chicago region.) Overall, from 1910 to 1970, there were 74 blast furnaces built and 94 scrapped in the former locations, while 63 were built and 248 were scrapped in other northern regions, the south and the west.

Previously, it was noted that plants in the slow growth, high transportation cost regions would have been at a disadvantage compet-ing with plants in the high growth, high demand regions, even if they were comprised solely of the largest new furnaces, because of the importance of transportation costs of raw materials. The same conclu-sion certainly would hold for existing, moderate to small sized furnaces in high transportation cost areas, with the conclusion that economic obsolescence was hastened in those areas (that is, the plants in those

regions were the ones scrapping the larger furnaces noted in Figure 9. 2).
Although transportation costs declined in relative importance during the
1950s and 1960s, the variance in this component of cost (from one to
two dollars per ton of iron produced) in the well-located steel centers
was still large enough to exert some influence on decisions to scrap.

COST FACTORS

In this section, the input-output data presented in earlier chapters
will be coupled with typical prices that prevailed in the mid-1960s to
indicate the relative economies that could be achieved by (1) building
relatively large new furnaces compared with smaller furnaces, and
(2) concentrating on processing materials more extensively rather than
building new furnaces. Both of these comparisons will be made under
conditions of strong and weak demand. These case studies are representative only for furnaces located in plants in the Great Lakes region.

Data on investment costs for new furnaces are scarce and subject
to large inaccuracies. This is because construction costs depend
strongly on regional factors and conditions in the plant in which the
furnace is built. Most engineering estimates of the cost of a new
furnace indicate that the power rule (with a value of two-thirds) is
applicable up to furnace sizes that have been built previously. David
Dilley believes that the power rule is the best rule of thumb that can
be employed for estimating costs prior to considering the economic
details of the plant in which the furnace is to be built.[*] Accordingly,
it was assumed that the power rule with a power value of two-thirds
applied to furnaces with hearth diameters in the 25- to 35-feet range.
The primary effort was concentrated on obtaining accurate estimates
of the construction costs of a furnace with a hearth diameter of 30
feet. Then using the relationship, Investment Cost = $A(Capacity)^{2/3}$,
where A is a constant, the costs of furnaces of other sizes were
estimated. C. Dryden and R. Furlow and G. Manners and G.
Meinhausen[9] were in reasonably close agreement that furnaces with
a hearth diameter of 30 feet (which, when operated on agglomerates
with fuel injection, are capable of producing 3,000 tons per day) cost
in the neighborhood of $20 million in the mid-1960s, including all
ancillary equipment. Based on this estimate and using the power rule,
it was calculated that a furnace with a hearth diameter of
- 25 feet cost $16, 030, 000, and
- 35. 4 feet cost $25, 440, 000.
All other capital costs involved in the estimates of coke plants and

---

[*]This information was received in a telephone conversation with
Dilley in the summer of 1970.

pelletizing facilities were assumed to be included in the prices of the inputs to the blast furnace process.

Comparing Costs of Large and Small Furnaces

Using approximations of the detailed relationships between capacity, furnace size, and materials quality that were developed in Chapter 7, it has been assumed that capacity is directly proportional to furnace hearth area, given the quality of the materials. Thus, one furnace with a hearth diameter of 35.4 feet has twice the capacity of a furnace with a hearth diameter of 25 feet. Additionally, it has been assumed that a furnace operating on Mesabi non-Bessemer ore with an iron content of 51.5 percent is capable of producing two net tons per day per square foot of hearth area, and a furnace operating on pellets with an iron content of 62 percent is capable of producing four net tons per day per square foot of hearth area.

From the estimates of labor requirements in Chapter 8, it was estimated roughly that the smaller furnace operating on natural ore requires a labor force of 34 men per 8 hour shift; and the larger furnace requires a labor force of 36 men (two additional men are required to handle the additional flue dust generated on the larger furnace). When these furnaces are operated on pellets, it was assumed that the smaller furnace needed 36 men and the larger one 42 men (because it is necessary to add a second taphole and casthouse).

In Chapter 7 it was estimated that each increase of one foot in the hearth diameter decreased the coke rate by 0.008 tons for blast furnaces with hearth diameters from 16 to 28 feet. The estimates in this section will be made with and without this saving, since some statistical doubt attaches to its magnitude and because this saving may not be applicable to furnaces larger than 30 feet.

From the 1963 Census of Manufactures and the Bureau of Labor Statistics, the average hourly wage including the employer's payroll tax has been approximated at $4. From the 1966 Minerals Yearbook, the price of coke has been approximated at $16 (this is the average value of one net ton of coke at iron and steel plants, rather than a market price). These approximations are presented in Table 9.5. Table 9.5 indicates that the superior efficiency of a blast furnace with a hearth diameter of 35.4 feet versus a furnace with a hearth diameter of 25 feet could have amounted to a saving in unit costs that ranges from $1.19, if pellets were used and no coke saving were achieved in the larger furnace, to $3.77, if natural ore were used and a coke saving of 0.080 net tons per net ton of iron were realized. It is worth noting that the saving in unit capital and labor costs would have been smaller if the furnaces were operated on agglomerates.

These estimates of the cost savings realized with a larger furnace in the mid-1960s are pertinent to periods of peak demand when plants

TABLE 9.5

Estimated Cost Advantage of a Large Blast Furnace in the Mid-1960s in the United States

	Natural Ore Burden Hearth Diameter		Pellet Burden Hearth Diameter	
	25 Feet	35.4 Feet	25 Feet	35.4 Feet
1. Furnace output: per day (net tons)	982	1,964	1,964	3,928
per year (net tons × 10^3)	358	717	717	1,434
2. Investment cost ($ × 10^6)	16.03	25.44	16.03	25.44
3. Labor force per shift (men)	34	36	36	42
4. Investment cost per annual ton (dollars)[a]	44.80	35.50	22.40	17.80
5. Man-hours per ton[b]	0.83	0.44	0.44	0.26
6. Prorata factor to convert investment cost per annual ton into annual depreciation and interest charges per ton[c]	0.1	0.1	0.1	0.1
7. Hourly wage rate (dollars)	4	4	4	4
8. Depreciation and interest charges per ton (dollars)[d]	4.48	3.55	2.24	1.78
9. Labor cost per ton (dollars)[e]	3.32	1.76	1.76	1.03
Subtotal	7.80	5.31	4.00	2.81
10. Capital and labor unit cost saving in larger furnace (dollars)		2.49		1.19
11. Possible excess unit coke cost in smaller furnace (dollars)[f]		1.28		1.28
12. Maximum unit cost saving in larger furnace (dollars)		3.77		2.47

[a] Derived by dividing (2) by per year furnace output.
[b] Derived by dividing (3) × 24 hours by per day furnace output.
[c] This factor assumes the investment cost is amortized over 20 years at 7-3/4 percent.
[d] Derived by multiplying (4) by (6).
[e] Derived by multiplying (5) by (7).
[f] Derived by multiplying the higher coke rate of 0.080 (= 0.008 × 10 feet) by $16.

Source: Compiled by the author.

decide to add additional capacity. They indicate that there would have been a definite advantage in choosing a large furnace over two smaller furnaces whenever the demand for iron supported a major increment in capacity.

When demand was not strong enough to support new capacity without abandoning or scrapping older and smaller furnaces, the cost figures indicate it would have been optimal to leave the blast furnace stock unchanged (see Table 9.6). Two possibilities were considered for both natural ore burdens (with an iron content of 51.5 percent) and pellet burdens (with an iron content of 62 percent); (1) replacing two medium-size furnaces (with hearth diameters of 25 feet) with a new large furnace (with a hearth diameter of 35.4 feet); and (2) replacing three small furnaces (with hearth diameters of 20.4 feet) with the same large furnace. In Table 9.6, the figures indicate that if unit coke savings of 0.008 ton were realized for each foot by which the new furnace's hearth diameter exceeded that of the smaller furnaces, it would have been advantageous to replace the smallest furnaces (with hearth diameters of 20.4 feet); and it would have been economical to replace the medium-size furnaces if pellets were used in the furnaces. If no unit coke saving were realized (or a much smaller unit coke saving were realized), then it would not have paid to replace either the small or medium size furnaces. In all cases, it was assumed that the cost of capital was 7-3/4 percent.

Comparing Cost Advantages of Various Capacity Increases

To indicate the relative economies that could have been achieved by more extensive materials processing as compared with constructing new furnaces, the following situation will be explored:

(1) the blast furnace plant was comprised of three furnaces, each 30 feet in hearth diameter;

(2) one furnace was operated on a 100 percent pellet burden; the other two used Mesabi non-Bessemer ore; and

(3) forecasts indicated a sufficient increment in demand to justify constructing a fourth furnace with a hearth diameter of 30 feet to be operated on natural ore, or operating a second existing furnace on a 100 percent pellet burden.

To estimate the unit costs associated with these alternatives, prices that prevailed in the mid-1960s and unit input requirements explored in other chapters will be employed. It will continue to be assumed that the price of coke was $16 a net ton and the hourly wage rate was $4. In the 1966 Annual Statistical Report of the American Iron and Steel Institute, the price of one gross ton of standard Mesabi non-Bessemer ore was reported to be $10.55, and the price of a gross ton

TABLE 9.6

Estimated Cost Advantage of Replacing Smaller Blast Furnaces with a Larger One

| | Natural Ore Burden | | | Pellet Burden | | |
| | Hearth Diameter | | | Hearth Diameter | | |
	20.4 Feet	25 Feet	35.4 Feet	20.4 Feet	25 Feet	35.4 Feet
1. Output per blast furnace day (net tons)[a]	655	982	1,964	1,309	1,964	3,928
2. Labor force per shift (men)[a]	32	34	36	34	36	42
3. Man-hours per ton[b]	1.17	0.83	0.44	0.62	0.44	0.26
4. Labor cost per ton (dollars)[c]	4.68	3.32	1.76	2.48	1.76	1.03
5. Unit labor cost disadvantage of smaller furnaces (dollars)	2.92	1.56	—	1.45	0.73	—
6. Depreciation and interest charges on large furnace per ton of annual capacity (dollars)[a]	—	—	3.55	—	—	1.78
Subtotal (5 + 6)	2.92	1.56	3.55	1.45	0.73	1.78
7. Possible excess unit coke cost in smaller furnaces (dollars)[d]	1.92	1.28	—	3.37	2.01	—
Total (5 + 6 + 7)	4.84	2.84	3.55	3.37	2.01	1.78
8. Maximum unit cost disadvantage of smaller furnaces	1.29	-0.71	—	1.59	0.23	—

[a] See Table 9.5.
[b] Derived by dividing (2) × 24 hours by (1).
[c] Derived by multiplying (3) by $4/hour.
[d] Derived by multiplying 0.008 by the difference in hearth diameters (35.4 − HD) by $16 per ton of coke.

Source: Compiled by the author.

175

of pellets (with an iron content of 62 percent) was reported to be $15.62. (These prices are quoted for ores "at rail of vessel" at lower Lake Erie ports.) The gross-ton prices are equivalent to net-ton prices of $9.42 and $13.95, respectively. Other sources indicate the price of flux was approximately $2 per net ton.[10]

The coke rate will be estimated on the basis of equation 6.6 (which is based—somewhat conservatively—on operating data supplied by company X); and the flux rate is based on equation 6.4a. The previous assumptions about blast furnace productivity and the required labor force will continue to be employed in this chapter—namely, that a furnace operating on standard ore is capable of producing two net tons per day per square foot of hearth area, and a furnace operating on pellets can produce at twice that rate. The blast furnace labor force will be assumed to be 36 men. Slightly more labor is needed to process flue dust when a furnace is operating on standard ore, but slightly more labor also is needed to handle the doubled volume of iron when a furnace is operating on pellets.

Table 9.7 indicates there would have been a sizable cost advantage to upgrading materials to achieve capacity increases. The estimated advantage of $7.60 per net ton of pig iron is large enough to show the superiority of pellets, even if the coke rate were reduced less substantially than indicated in Table 9.7. This large advantage provides a margin of "safety" for the likelihood that natural ore sold for a discount, while pellets sold for a premium, compared with reported prices. In fact, the operating cost advantage of pellets of $3.77 indicates the optimal policy would have been to expand pelletizing capacity and/or purchase pellets from other sources, even if this policy caused some existing blast furnaces to be shut down (should the choice have had to be made between natural ore and pellets). Of course, the latter conclusion is unlikely to hold if pellets are compared with sinter or high grade foreign ore.

In summary, the data indicate that while large blast furnaces were more efficient than smaller furnaces, their edge in efficiency was reduced when agglomerates of high quality were used (for example, pellets) in place of standard domestic ore. Until a plant experienced a large increase in demand, however, it would not have been to its advantage to construct a large new furnace (a hearth diameter of 35 feet) unless (1) it was already operating most furnaces on high quality agglomerates (where the maximum cost saving could be achieved) and (2) it had a number of smaller furnaces (three furnaces with hearth diameters less than 20 feet) that could be shut down—and this latter requirement assumes that the large new furnace could be operated with a significantly lower coke rate. These general conclusions would hold even more convincingly when the interest rate goes higher than the 7-3/4 percent assumed in the preceding three tables or when the power value was not as low as two-thirds. Furthermore, even great inaccuracies in the estimated unit labor cost would be unlikely to alter these conclusions.

TABLE 9.7

Estimated Unit Cost Advantage of Expanding Capacity by Using More Pellets Rather Than Building a New Furnace and Using Natural Ore

	Natural Ore (iron content = 51.5%)	Pellets (iron content = 62%)
1. Man-hours per net ton of iron produced	0.62	0.31
2. Burden rate [a]	1.82	1.51
3. Coke rate	0.90	0.52
4. Flux rate	0.43	0.23
5. Hourly wage (dollars)	4.00	4.00
6. Price of iron-bearing materials (dollars)[b]	9.42	13.95
7. Price of coke (dollars)[b]	16.00	16.00
8. Price of flux (dollars)[b]	2.00	2.00
9. Unit labor cost (dollars)	2.48	1.24
10. Unit ore cost (dollars)	17.22	21.17
11. Unit coke cost (dollars)	14.40	8.32
12. Unit flux cost (dollars)[b]	0.86	0.46
Unit operating costs	34.96	31.19
13. Depreciation and interest charges per net ton of iron (dollars)	3.88	—
Unit total cost[b]	38.84	31.19

Assumes pig iron is 94 percent iron by weight.
Per net ton.
Note: Table derived from data discussed in text.

It is improbable that many U.S. plants fulfilled the conditions necessary to make large new furnaces economically attractive. Hence, the U.S. had less economic incentive than Japan, the Soviet Union, and other countries to adopt large, new furnaces.

NOTES

1. William T. Hogan, S.J., Productivity in the Blast Furnace and Open-Hearth Segments of the Steel Industry: 1920-1946 (New York: Fordham University Press, 1950), p. 39.

2. Because the hearth diameter was used as an indicator of the size of new blast furnaces, it was necessary to estimate hearth diameters from bosh diameters prior to 1935. The Directory of Iron and Steel Works in America (New York: American Iron and Steel Institute, 1935) was the first issue of that series to report the hearth diameters of U.S. blast furnaces. Using the 1935 Directory, it was possible to make accurate estimates of hearth diameters back to 1914. These estimates indicated that the average difference between bosh diameter and hearth diameter for new furnaces was 4.8 feet in 1914-15, 4.3 feet in 1916-20, and 3.5 feet in 1921-25. According to a sample of new blast furnaces reported by J.E. Johnson in Blast Furnace Construction in America (New York: McGraw-Hill, 1917), the most common difference between these diameters in 1900 was six feet. Hence, it was assumed that this differential was 6 feet in 1900-05, 5.25 feet in 1906-10, and 4.75 feet in 1911-13. Sufficient regularity in the dimensions of new furnaces had been established by 1900 to make these estimates reasonable.

3. See Tables 7.3 and 7.5.

4. All issues of the American Iron and Steel Institute Annual Statistical Report (New York: AISI, from 1914 to 1959) reported total capacity. Because of the significant improvement in the quality of the iron-bearing materials after 1955, this technique will not be applied to the last five reported capacity figures.

5. David Dilley and David McBride of United States Steel and J. Fischley of Republic Steel each indicated in separate interviews during 1970 that iron capacity requirements were derived from expected steel products sales and that an attempt is made to maintain a balance between the various stages of production in integrated plants.

6. Hogan, op. cit., pp. 38-39.

7. William Collison, formerly superintendent of blast furnaces at the Great Lakes Plant of National Steel and currently with Arthur G. McKee, admitted to holding this belief until the theory was disproved (1970 interview).

8. George Stigler, "The Economies of Scale," The Journal of Law and Economics 1 (October 1958).

9. C. Dryden and R. Furlow, Chemical Engineering Costs (Columbus: Engineering Experiment Station, 1966), p. 71; G. Manners, The Changing World Market for Iron Ore 1950-1980: An Economic Geography (Baltimore: The Johns Hopkins Press for Resources for the Future, 1971), p. 37; and G. Meinhausen, "Iron and Steel Works — Maximum Capacity, State of Planning and Chances for Development," Stahl und Eisen 90 (19 February 1970): 156-57.

10. H. M. Graff and S. C. Bouwer, "Economics of Raw Materials Preparation for the Blast Furnace," Journal of Metals 17 (April 1965): table II, p. 391.

The blast furnace sector of the steel industry has been somewhat unusual because the basic technique of production has been the blast furnace for centuries. Although many modifications and refinements have been made in this device, the fundamental principles of trans-forming iron ore into pig iron have remained the same. A number of alternative methods of transforming iron ore into iron or steel have been developed during the last 40 years (they generally are referred to as direct reduction techniques), and a few of these are in commer-cial operation on a limited basis and in special circumstances. But the blast furnace has remained the basic initial capital instrument used in iron and steelmaking. This common denominator facilitated the analysis of the preceding chapters using the model developed in Chapter 3 because the basic activity was unchanged, even though the work done in this activity has diminished during the last two decades.

The preeminence of the blast furnace as a source of iron in the entire 1900-1970 period raises the question of whether the performance of the blast furnace activity was inferior to all manufacturing. That is, is the longevity of the blast furnace indicative of technological back-wardness (for whatever reason) in this activity? To answer this ques-tion, it is necessary to develop some measure of performance, both in the blast furnace activity and all manufacturing.

The wholesale price index will be used as a yardstick for the economic performance of all manufacturing industries, against which the performance of the blast furnace sector may be contrasted, under the assumption that variations in profit margins have little effect on the price of manufactured commodities over the long run and, hence, that this index measures the net effect of improvements in productivity (of all inputs through technological change, scale, and such) and changes in factor prices over time. There are some conceptual diffi-culties in using the wholesale price index as an indicator of economic performance in all manufacturing. Perhaps the greatest of these is the fact that the representative sample of industrial commodities underlying this index has changed drastically during the twentieth century (unlike pig iron), raising some doubts about the internal consistence of a long series of these prices or even the meaning of this series. But it is unlikely that a better indicator of the performance of the manufacturing sector could be found.

The relative performance of the blast furnace sector can be measured by either a composite price series for pig iron, deflated by the wholesale price index, or by value added per ton of iron produced, deflated by the wholesale price index. Neither measure is completely satisfactory. The total measure (that is, the deflated composite price series) requires that profit margins be constant over broad periods of time.* Additionally, it may distort the performance of the blast furnace sector since it is affected by the prices and qualities of materials, over which blast furnace managers may have little control. The net measure (deflated value added per ton of iron produced) also requires that profit margins remain approximately constant and should be adjusted for variations in the work done in the blast furnace sector over time. There are two measures of value added that can be used. The first is the conventional measure of the value of capital and labor services. The second adds the value of energy consumed to the first. The second measure appears to be more valid than the first since energy, as well as the services of capital and labor, is used up in the process of transforming materials into higher level products. The problem with the more complete measure of value added is that coke is the primary energy input, and coking coal is used almost solely by the blast furnace sector; but other industries derive their energy from commonly used fuels such as steam coal and natural gas. Hence, if the price of coking coal does not reflect patterns faced in other industries, the relative performance of the blast furnace sector will be distorted.

The changes in the work done in the blast furnace sector were substantial between 1955 and 1970. But it is by no means clear how one should proceed to adjust deflated value added per ton of iron produced to reflect the diminished work load in this sector if improvements in materials had an uneven impact on capital, labor, and energy. According to the analyses in preceding chapters, improved iron-bearing materials permitted generally uniform reductions in the unit capital and direct labor inputs in the blast furnace sector but allowed greater reductions in the unit capital input than the unit total labor input (because indirect labor may not, and distributive labor certainly is not, reduced in proportion to the number of furnaces that may be taken out of blast). Proportionately smaller reductions in the coke rate than

---

*More precisely, the profit earned should reflect the competitive return to capital and entrepreneurship but should not include elements of monopoly profit.

in the unit capital input were measured. Hence, improved materials appear to have reduced the capital requirement more than the labor requirement, and the labor requirement more than the coke rate.

Since neither measure of blast furnace performance is totally satisfactory, the total measure of performance will be discussed first. Subsequently, the net (value added) measures will be reviewed.

CATEGORIES OF INPUTS AND THEIR PROPORTIONS OF TOTAL REVENUE

In the assessment of the performance of the blast furnace sector, it is convenient to consider the input costs as the product of a price and a quantity. It is possible to achieve this breakdown for ores and agglomerates (that is, all iron-bearing inputs except scrap), coke, and hourly workers and salaried employees. In Chapter 9 a crude time series of the unit capacity cost of new and rebuilt furnaces was developed, but this series was based on data from a small sample and, more important, does not measure the cost of capital services in the economic or accounting sense (that is, interest and deprecia-tion charges).

Thus, the inputs in this chapter will be divided into six categories: ores and delivered agglomerates (chiefly pellets), coke, other materials and energy sources (a heterogeneous category), hourly workers, salaried employees, and overhead and profit (a heterogeneous category). From Table 10.1 it can be seen that the four inputs that are susceptible to "price times quantity" breakdowns accounted for as much as $83\frac{1}{2}$ percent of total revenue (in 1914) and no less than $65\frac{1}{2}$ percent of total revenue (in 1963). Although some sizable fluctuations in these costs proportions are evident, no discernable pattern of change can be detected, except for wages—which fell—and "other materials and energy costs," which rose. The figures indicate that the proportions of these inputs that can be decomposed into "price times quantity" (other than hourly labor), when measured in value terms, have tended to remain constant in the long run.

FACTOR PRICES

Since most blast furnace plants are components of integrated steel firms, the prices "paid" for coke, iron-bearing materials, and pig iron are generally transfer prices rather than market prices. This has been particularly true since the late 1920s. While long term changes in these prices probably reflected changes in the average costs of production and shipment with reasonably accuracy, short term changes could have been tailored to suit managerial preferences. This observation applies with particular emphasis to the iron-bearing

TABLE 10.1

Cost Proportions, as a Percentage of Total Revenue

Year	Ores and Delivered Agglomerates (1)	Coke, Coal, Charcoal (2)	Other Materials and Energy Costs[a] (3)	Wages (4)	Salaries (5)	Overhead and Profit[b] (6)
1904	43.5	26.5	7.2	8.2	1.3	13.3
1909	48.0	27.1	7.0	6.2	1.7	10.0
1914	47.5	26.9	8.9	7.2	1.9	7.6
1919	39.1	29.2	10.5	9.4	1.7	10.1
1929	45.9	24.5	8.7	5.5	0.9	14.5
1937	44.5	25.8	10.7	5.6	0.8	12.6
1939	45.5	26.1	12.6	5.2	1.0	9.6
1947	30.0	36.0	14.9	5.4	1.1	12.6
1954	34.7	29.2	13.5	4.8	1.2	16.6
1958	36.0	26.9	13.8	3.9	(1.2)[c]	18.2
1963	37.1	23.2	14.8	4.0	(1.2)	19.7
1970	38.4	30.4	15.3	3.8	(1.2)	10.9

[a]Includes flux, scrap, natural gas, fuel oil, tar, and coke breeze used in sintering and other supplies.
[b]Includes interest, state and local taxes, depreciation and amortization, fixed employment costs (such as social security contributions, pensions, sick pay, and other fringe benefits), dividends, retained earnings, and federal income tax.
[c]Salaries for the period 1958-70 were guessed to be equivalent to their 1954 proportion. Comprehensive data pertaining to this cost proportion were not available after 1954.

Sources: Census of Manufactures (Washington, D.C.: Bureau of the Census, all issues from 1904 to the present); and M.G. Boylan, "The Economics of Changes in the Scale of Production in the U.S. Iron and Steel Industry from 1910 to 1970" (Case Western Reserve University, unpublished Ph.D. thesis, 1973), Appendix C.

FIGURE 10.1

Cost Proportions, as a Percentage of Total Revenue

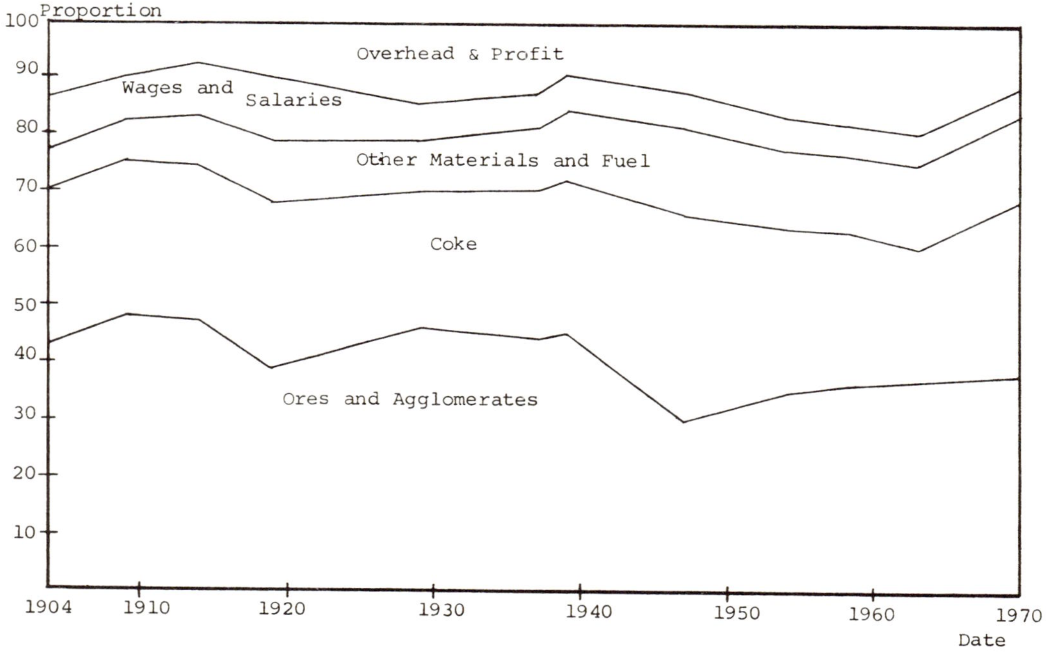

Source: Table 10.1

184

materials, not only because the major steel firms traditionally have controlled their sources of these materials (including sources in foreign countries)[*] but also because the alternative uses for these materials are practically nonexistent. In contrast, most of the major steel firms have been dependent to some extent on private suppliers of coking coal because of insufficient capacity in their captive mines. There is also an opportunity cost to using coking coal in the blast furnace sector since it was at least a partial substitute for steam coal. Hence, the price of coking coal could not fall without limit in relation to the price of steam coal, because the resulting increase in demand would have pushed the price back up to a certain minimum level.

From 1904 to 1947, the price of iron-bearing materials exhibited greater stability than the wholesale price index during inflationary and deflationary periods, with the result that ore was relatively cheaper during inflationary periods, and the other way around (see Table 10.2 and Figure 10.2). From 1947 to 1954, however, the price of ore almost doubled while the wholesale price index rose by only 15 percent, reflecting increased efforts to upgrade ore at domestic mines (the proportion of domestic ore concentrated rose from 23 percent to 36 percent), an increase in the proportion of ore imported (from 9 percent to 18 percent), and the increasing scarcity of ore of normal quality (that is, a natural iron content of 51.5 percent). It is possible that the latter reason was a large contributing factor to higher ore prices because increased prices are a good rationing device, and they provided the owners of the major mines with funds with which to develop both alternative sources of ore in foreign countries and pelletizing facilities in the Lake Superior region. Furthermore, despite increases in imports and a rapid expansion of pelletizing facilities, the rise in the price of iron-bearing materials relative to the wholesale price index was reduced from 1954 to 1958 and slightly reversed during the 1960s.

While changes in the prices of iron-bearing materials did not seriously disadvantage the blast furnace sector relative to all manufacturing, Table 10.3 and Figure 10.3 indicate that the changes in wage rates and coke prices would have contributed to rapid increases in the average total cost of pig iron if it had not been possible to lower the consumption of labor and coke relative to the production of pig iron. The wage rate relative to the composite price for iron-bearing materials rose rapidly from 1909 to 1970, from an index of 41 to 273. The price of coke broadly followed the same pattern as the price of iron-bearing materials but exhibited greater sensitivity to general economic conditions (as would be expected). During the

---

[*]Foreign nations can, of course, influence the delivered prices of materials shipped from their mines.

TABLE 10. 2

Composite Price Indices for the Iron-Bearing Materials and the
Wholesale Price Index for All Manufactured Commodities
(Index: 1929 = 100)

| Year | Delivered Price of Ore and Agglomerates | | Wholesale Price Index | Deflated Delivered Prices of Ore and Agglomerates (4) |
	Per Gross Ton (1)	Adjusted (2)	(3)	(2) ÷ (3)
1904	72	70	63	111
1909	83	82	71	115
1914	77	77	71	108
1919	120	120	146	82
1929	100	100	100	100
1937	101	101	91	111
1939	103	103	81	127
1947	119	120	155	78
1954	237	235	179	131
1958	287	273	193	141
1963	290	266	193	138
1970	332	300	228	132

Sources: Column 1, U.S. Census of Manufactures (Washington,
D.C.: Bureau of the Census, all issues from 1904 to 1963). 1947 and
1970 were estimated from various sources. See M.G. Boylan, "The
Economics of Changes in the Scale of Production in the U.S. Iron and
Steel Industry from 1910 to 1970" (Case Western Reserve University,
unpublished Ph.D. thesis, 1973), Appendix C. Column 2, derived
from (1) by adjusting the price per gross ton for annual variations in
the natural iron content. The natural iron content of Lake Superior ore
was used to adjust 1904-47. Thereafter, estimates of the natural iron
content of all ores and agglomerates delivered to blast furnace plants
and sinter plants were employed. Column 3, U.S. Bureau of Labor
Statistics.

FIGURE 10.2

Comparison of the Composite Price Index for Iron-Bearing
Materials with the Wholesale Price Index (1929=100)

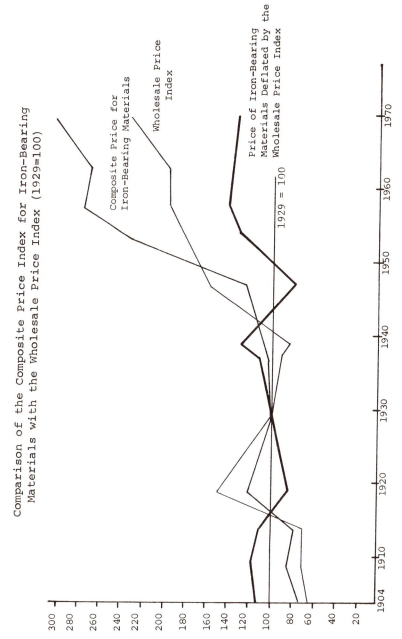

Source: Table 10.2

187

TABLE 10.3

Input Prices
(index: 1929 = 100)

| Year | Input Prices | | | Input Prices Relative to the Composite Price for Ore and Agglomerates | |
| | Labor | Coke | Ore and Agglom-erates | Labor | Coke |
	(1)	(2)	(3)	(4)	(5)
1904	33	65	70	47	93
1909	34	73	82	41	89
1914	39	70	77	51	91
1919	97	146	170	81	122
1929	100	100	100	100	100
1937	139	109	101	138	108
1939	144	108	103	140	105
1947	267	253	120	223	211
1954	420	357	235	179	152
1958	566	411	273	207	151
1963	665	396	266	250	149
1970	818	614	300	273	205

Sources: Column 1, 1904-63, Table 8.1; 1970 is the average hourly wage in basic steel. Column 2, 1904-54, Census of Manufactures (Washington, D.C.: Bureau of the Census, all issues from 1904 to 1954); 1958-70, U.S. Bureau of Mines, Minerals Yearbook (Washington, D.C.: U.S. Department of the Interior, 1958-70). Column 3, Table 10.2. Columns 4 and 5, derived by dividing (1) and (2), respectively, by (3) and multiplying by 100.

FIGURE 10.3

Wage Rates in the Blast Furnace Sector and the Price of
Coke, Relative to the Composite Price of Iron-Bearing
Materials(Index: 1929=100)

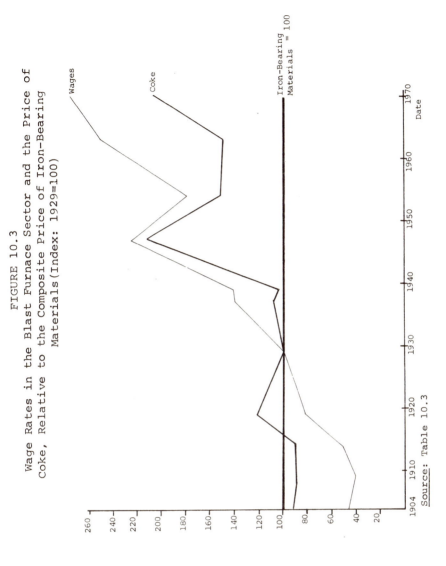

Source: Table 10.3

189

three subperiods 1904-19, 1919-29, and 1929-54, the relative increase or decrease in coke prices was almost exactly 85 percent greater than the corresponding change in the price of iron-bearing materials; and from 1954 to 1970 it was 160 percent greater. From 1904 to 1970, the price of coke rose by more than twice the increase in the price of iron-bearing materials.

Since coke is directly consumed (instead of being used to generate electricity, for example), the rise in the price of this important input after World War II certainly disadvantaged the blast furnace sector, in comparison with all manufacturing. The rise in wage rates, on the other hand, was in line with similar increases elsewhere.

INPUT-OUTPUT RATIOS

The real gains made by the blast furnace sector in lowering the cost of pig iron are measured by the input-output ratios reported in Table 10.4. From 1904 to 1954, the indicated reductions may be considered the result of technological improvements and increases in scale within the blast furnace sector. Over this half-century span of time, the unit labor requirement was reduced 83 percent, the coke rate was reduced 21 percent, the number of blast furnaces (of a given size) required to produce a given tonnage of iron was lowered by 17 percent,[*] and the unit quantity of iron-bearing materials consumed (of constant iron content) fell 7 percent as a result of reductions in flue dust and casting losses. The magnitude of these improvements strongly paralleled the magnitude of increases in the factor prices previously noted and partially offset those price increases.

From 1954 to 1970, the reductions in labor, coke, and utilized blast furnace capital must be considered to be at least as much the result of the improvements made in iron-bearing materials as the result of further gains in scale and technology in the blast furnace sector. This observation is particularly true for capital but less correct for coke and labor. Over this shorter period of 16 years, the

---

[*]Whether this represents an equal reduction in the consumption of capital services is difficult to estimate. Blast furnaces of a given size represented "more capital" over time as the ancillary equipment became more sophisticated. On the other hand, if the power rule is applicable to increases in the size of blast furnaces with an exponent that is less than one, then larger furnaces represented "less capital" per capacity unit than smaller furnaces, offsetting to some extent the increased cost of technological improvements. Finally, the greater durability of furnaces over time also reduced the consumption of capital services per capacity unit.

TABLE 10.4

Consumption of Labor, Coke, Iron-Bearing Materials, and Utilized
Blast Furnace Physical Capital per Ton of Iron Produced
(index: 1929 = 100)

Year	Labor (1)	Coke (2)	Iron-Bearing Materials (3)	Utilized Blast Furnace Physical Capital (4)
1904	353	122	103	—
1909	284	122	104	112
1914	252	114	103	—
1919	252	112	101	106
1929	100	100	100	100
1937	84	99	97	—
1939	63	97	98	102
1947	66	104	98	—
1954	59	96	97	93
1958	42	88	96	—
1963	34	74	94	69
1970	29	69	95	65

Sources: Column 1, Table 8.2. Column 2, Table 6.1. Column 3, Table 5.2, adjusted for differences in natural iron content; that is, this series measures the reduction in amount of iron-bearing materials wasted. Column 4, Table 7.7.

following reductions were achieved: labor—50 percent, coke—28 percent, capital—30 percent, and iron-bearing materials—2 percent.

UNIT COSTS RELATIVE TO THE COMPOSITE PRICE
OF IRON-BEARING MATERIALS

To obtain an impression of the pressure exerted on the composite price for pig iron products relative to the price of iron-bearing materials, Table 10.5 and Figures 10.4, 10.5, and 10.6 were constructed. Figure 10.4 illustrates that the broad trend in the unit cost of coke has been an increase relative to the price (and unit cost) of iron-bearing materials. But more important, large relative increases in this cost during the World War I and World War II years were reversed almost completely in the following years. The broad trend in the unit labor cost (relative to the price of iron-bearing materials) has been a modest

191

TABLE 10.5

Composite Price of Pig Iron and the Unit Costs of Iron-Bearing Materials, Coke, Other Materials and Fuels, Labor, Salaried Personnel, and Overhead and Profit, Deflated by the Composite Price of Iron-Bearing Materials
(index: 1929 = 100)

Year	Iron-Bearing Materials (1)	All Other Inputs (2)	Coke (3)	Other Materials and Fuels[a] (4)	Labor (5)	Salaried Personnel (6)	Overhead and Profit[b] (7)	Composite Price of Pig Iron (8)
1904	103	117	113	89	166	160	100	113
1909	104	101	109	79	118	198	71	102
1914	103	96	104	100	127	210	52	99
1919	101	142	137	152	203	234	85	124
1929	100	100	100	100	100	100	100	100
1937	97	105	105	126	86	93	86	102
1939	98	96	102	138	88	103	62	97
1947	98	180	219	141	147	188	129	155
1954	97	147	146	104	106	168	141	127
1958	96	147	133	107	87	167	153	127
1963	94	138	110	110	85	158	158	120
1970	95	136	141	121	79	155	86	118

[a]Includes flux, scrap, natural gas, fuel oil, tar, and coke breeze used in sintering and other supplies.

[b]Includes interest, depreciation and amortization, dividends and retained earning (allocated to the blast furnace department by firms reporting to the Census of Manufactures), federal income tax, state and local taxes, and fixed employment costs (such as social security contributions, pensions, sick pay, and other fringe benefits.)

Sources: Census of Manufactures (Washington, D.C.: U.S. Department of Commerce, Bureau of the Census, all issues from 1904 to present) and M.G. Boylan, "The Economics of Changes in the Scale of Production in the U.S. Iron and Steel Industry from 1910 to 1970" (Case Western Reserve University, unpublished Ph.D. thesis, 1973), Appendix C.

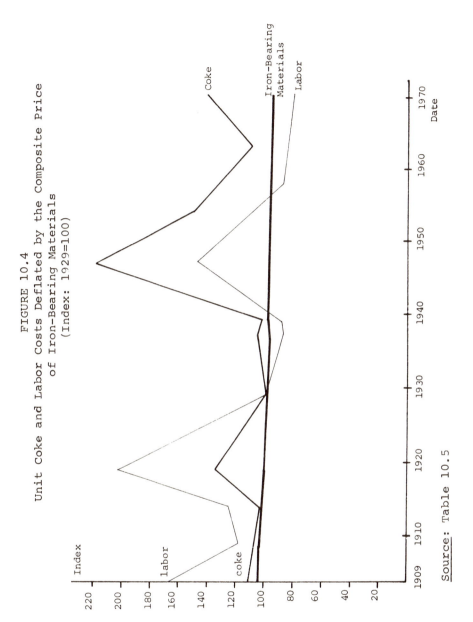

FIGURE 10.4

Unit Coke and Labor Costs Deflated by the Composite Price
of Iron-Bearing Materials

(Index: 1929=100)

Source: Table 10.5

193

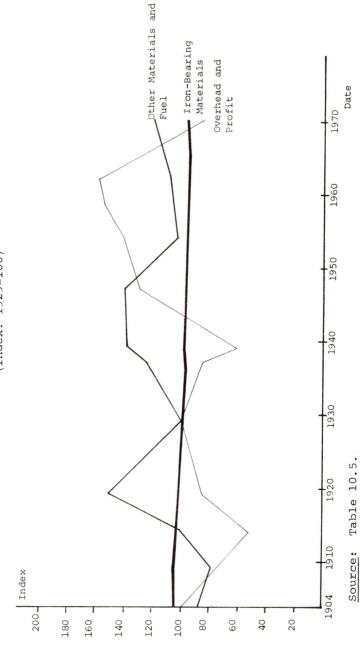

FIGURE 10.5

Unit Costs for Other Materials and Fuel and Overhead and Profit
Deflated by the Composite Price of Iron-Bearing Materials
(Index: 1929=100)

Source: Table 10.5.

194

FIGURE 10.6

Unit Costs of the Iron-Bearing Materials and all Other
Inputs, and the Composite Price of Pig Iron, Deflated by the
Composite Price of Iron-Bearing Materials
(Index: 1929=100)

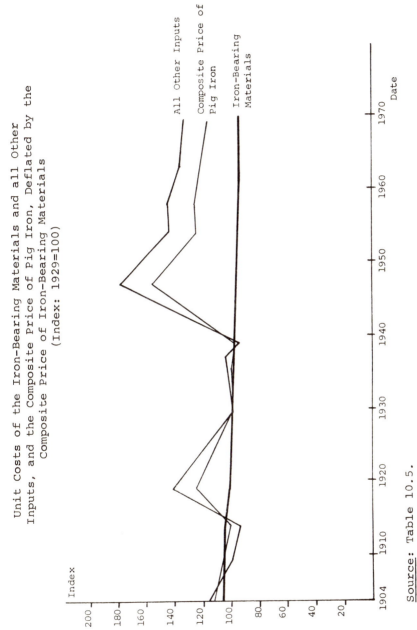

Source: Table 10.5.

195

decline. This cost also exhibited significant gains relative to the price of the iron-bearing materials during the two major wars, but these gains were completely reversed within a decade.

Figure 10.5 illustrates that during 1909-19 and during the last three decades the unit cost of other materials and fuel tended to increase the composite price of pig iron relative to the price of iron-bearing materials. This cost also was inflated during the two major wars but was reversed almost completely within the following decade in each postwar period. Overhead and profit per ton of iron was increased from 1904-39 to 1947-70, primarily because fixed employment costs, state and local taxes, and federal income tax were increased greatly per ton of iron. Overhead and profit per ton of iron fell during the recession/depression years of 1914, 1939, and 1970, indicating that unit profit fell more rapidly than unit overhead cost rose.

Figure 10.6 presents a summary measure of the unit cost of all inputs except the iron-bearing materials, relative to the price of iron-bearing materials. This series indicates that during the first four decades of this century only the inflation caused by World War I forced the composite price of pig iron products significantly above the composite price for iron-bearing materials, relative to their respective values in 1929. After World War II, however, the rapid increase in the prices of most of the other inputs compared with a modest 20 percent increase in the price of iron-bearing materials forced the composite price of pig iron to rise by 55 percent more than the price of iron-bearing materials. Even the subsequent rapid increase in the price of iron-bearing materials and the significant improvements in the productivities of the other inputs did not reverse completely the relative increase in the price of pig iron because the prices of these other inputs continued to rise rapidly.

SUMMARY MEASURES OF THE PERFORMANCE
OF THE BLAST FURNACE SECTOR

The composite price of pig iron and the two measures of values added per ton of iron relative to the wholesale price index tell basically the same story (see Table 10.6 and Figures 10.7 and 10.8). From 1904 to 1929, modest improvement in the performance of the blast furnace sector were achieved relative to all manufacturing. Major increases in the scale of furnaces and significant advances in mechanization paved the way for these advances.

From 1929 to 1947 the relative performance of the blast furnace sector worsened moderately. Further advances in scale during this period were small; the major gains were made in the area of modernizing the existing blast furnace stock. From 1947 to 1958 the performance

TABLE 10.6

Composite Price of Pig Iron, Value Added per Ton of Iron, and Value Added and Energy Cost per Ton of Iron, Deflated by the Wholesale Price Index

(index: 1929 = 100)

Year	Composite Price of Pig Iron (1)	Value Added per Ton of Iron (2)	Value Added + Energy Cost per Ton of Iron (3)	Wholesale Price Index (4)	Deflated by the WPI		
					Composite Price of Pig Iron (5)	Value Added per Ton of Iron (6)	Value Added + Energy Cost per Ton of Iron (7)
1904	79	85	85	63	125	135	136
1909	86	74	85	71	121	104	120
1914	77	61	74	71	108	86	104
1919	149	150	165	146	102	103	113
1929	100	100	100	100	100	100	100
1937	100	91	99	91	110	100	109
1939	94	71	87	81	116	88	107
1947	176	161	214	155	114	104	138
1954	283	303	322	179	158	169	180
1958	346	384	384	193	179	199	199
1963	319	378	338	193	165	196	175
1970	354	269	360	228	155	118	158

Sources: Columns 1, 2, and 3, U.S. Census of Manufactures and M.G. Boylan, "The Economics of Changes in the Scale of Production in the U.S. Iron and Steel Industry from 1900 to 1970 (Case Western Reserve University, unpublished Ph.D. thesis, 1973), Appendix C. Column 4, U.S. Bureau of Labor Statistics. Columns 5, 6, and 7, derived from the first four columns.

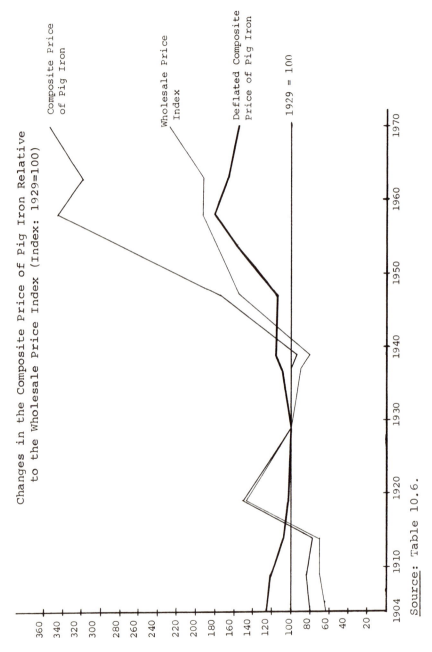

FIGURE 10.7

Changes in the Composite Price of Pig Iron Relative
to the Wholesale Price Index (Index: 1929=100)

Composite Price
of Pig Iron

Wholesale Price
Index

Deflated Composite
Price of Pig Iron

1929 = 100

Source: Table 10.6.

198

FIGURE 10.8

Value Added per Ton of Iron and Value Added Plus Energy Cost per Ton
of Iron, Deflated by the Wholesale Price Index

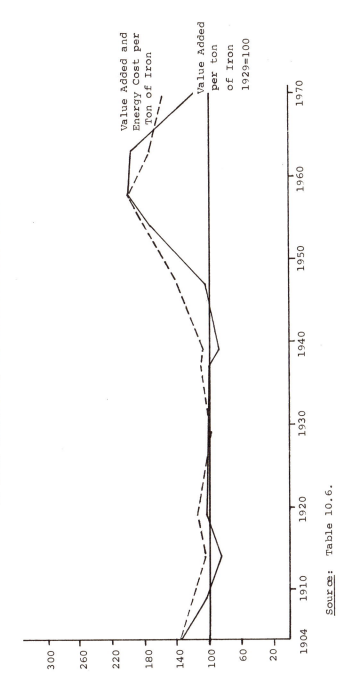

Source: Table 10.6.

of the blast furnace sector lagged significantly behind all manufacturing. The reasons were the rapid increases in the prices of the inputs used in this sector. The composite price of ores and agglomerates increased 125 percent, the wages of hourly labor rose 110 percent, and the price of coke increased 65 percent (having already increased 135 percent from 1939-47). Additionally, the blast furnace stock had been used heavily from 1941 to 1957 and required an unusually high level of maintenance and repair expenditures.

In the final 12 years of this study, the effect of improved materials and further gains in operating technology were felt as the blast furnace sector made gains in comparison with all manufacturing. With respect to the value added measures of performance, these gains were substantial—as would be expected, since the work done in this activity was diminished. On the other hand, the composite price of pig iron, deflated by the wholesale price of pig iron, measures not only the performance of the blast furnace sector but also the performance of earlier stages of production as well. Since the work done in the ore mining and preparation activities was expanded, the deflated composite price of pig iron would be expected to indicate smaller gains. This measure yields a 13 percent gain in the performance of the blast furnace sector from 1958 to 1970, while value added and energy cost per ton of iron fell 21 percent in the same period relative to the wholesale price index. (This latter series is dominated by coke. A 50 percent increase in the price of coke—concentrated in the last few years of this period--reduced the performance of the blast furnace sector according to this measure.) Value added per ton of iron fell 41 percent from 1958 to 1970 relative to the wholesale price index. This measure clearly overstates the gains made in the blast furnace sector because the profit component of value added was unusually low in 1970.

In summary, the blast furnace sector outperformed (or performed on par with) all manufacturing during the twentieth century, except for the 1947-58 period, according to the three measures developed in this chapter. Despite the fact that the major technique of production remained the same, this sector does not appear to have been unduly laggard in adopting performance improving innovations.

11

THE MODEL

The discussion in Chapter 1 and the review in Chapter 2 indicated, respectively, the limited applicability of the concept of economies of scale and the practical difficulties facing empirical investigations of such economies. Past investigations into economies of scale have foundered either because the long run average cost curve—a static cost curve—was not well suited to utilize the available data, because the prespecified production function reflected too many built-in assumptions about the nature of the true production processes, or because the models and data were too aggregate to uncover the desired relationships. To avoid the latter problem, a production model was proposed in Chapter 3 for capital-dominated and/or materials-dominated processes to investigate economies of scale at the subplant (activity) level to recognize the heterogeneity of activities.

The emphasis of this model was on the measurable, time-related changes in basic production processes, which included not only increases in scale but also related technical improvements (in the quality of the materials, in the design of the capital equipment, and such). Although this approach may not be as satisfying in the abstract as the traditional static models of scale (that is, the long run average cost curve or prespecified production functions), it has the merit of focusing attention on what is measurable while explicitly dealing with the possibility that increases in the scale of operations within any particular activity may be accompanied by changes in the work done by that activity (in the value added by that activity) due to changes in the quality of materials or output, or both. The model emphasizes in a general way the integration of each activity into a sequence of activities representing an industrial plant; a change originating in any particular activity (including an increase in the scale of operations in that activity) should be evaluated not only for its effect on cost within that activity but also for its effect on costs, capital requirements, and product quality in other activities connected with it. The (financial) evaluation of such changes in the parameters of the production model are best undertaken in a capital budgeting format.

The usual approach in studies of the effects of increased scale and process innovations on operating costs is to measure the reduction in the unit costs of the resources that are consumed completely during the production period: energy and the services of capital and labor. This technique of measuring reductions in value added per unit of output is correct as long as materials play only a passive role in the production process; that is, they are homogeneous in quality and are transformed into a uniformly "higher state" within the activity being investigated. This approach generally is not valid in empirical studies attempting to explain changes in production over broad periods of time and in studies of extractive industries where the materials are raw materials whose quality depends on past acts of nature. In the latter case the quality of the materials plays a central role in the production processes, and the adverse effect of improved materials quality on materials cost must be considered in conjunction with the beneficial effects of the improved quality of the materials on capital, labor, and energy costs. Furthermore, the study of production processes in extractive industries at the activity level of aggregation over a period of time is made even more difficult by the fact that, as the quality of the materials changes, it is often not possible to preserve comparability between the average costs of different periods by moving to a sequence of activities, because extractive activities typically are found at the beginning of a vertically integrated sequence of activities.

In this connection, it may be recalled that the quality (and sources) of iron-bearing materials was reasonably homogeneous during the 1910-45 period, that gradual changes in quality occurred from 1945 to 1955, and that dramatic changes took place from 1955 to 1970—changes that required explicit treatment in the analysis.

In a broader sense, the model developed in this book represents an appeal to the economics profession concerning the methodology of applied microeconomics. If we are ever going to be able to explain the evolution of plants, firms, and industries in a basically free enterprise, free choice economy, then it will be necessary to familiarize ourselves with the rudiments of production technology, to work at low (as well as high) levels of aggregation and to incorporate the effects of time explicitly into the analysis.

THE BLAST FURNACE INDUSTRY

The Past

The secondary purpose of this study was to explain increases in the scale of production in the blast furnace sector of the steel industry since the beginning of this century. The findings indicated that managers preferred larger furnaces throughout this period. Whenever new furnaces were added to a given plant, the sizes of the new furnaces were at least the same and frequently larger than the sizes of existing furnaces. Also, many furnaces were rebuilt a number of times to larger dimensions after their initial construction. Most furnaces scrapped during this period were smaller than the remaining furnaces in the same plant or other plants under common management within the same region.

Although it is likely that larger new furnaces were more efficient than smaller new furnaces throughout this century, only labor cost savings definitely were achieved. Data relating to capacity costs were insufficient to permit a determination concerning the behavior of this component of total cost at a point in time as blast furnace size was increased. In fact, the variance in the size of new furnaces constructed at a point in time was too small after 1920 to make an accurate determination of the effective savings in capital costs, even if data on these costs had been complete. Finally, although the data on company X's blast furnace operations during 1953-70 suggested that larger furnaces achieved modest savings in coke costs, these savings could have been due to a number of factors not included in the analysis.

Since labor costs accounted at most for 10 percent of total cost (and no more than 5 percent during the last two decades), reductions in this component of total cost had little effect on the total, saving perhaps 2 percent of total cost and 4 percent of value added (including energy) for a doubling of furnace size. Even if the "two-thirds rule" applied to capital costs (it is by no means clear that it did), the savings in capital cost by building one large rather than two small furnaces would have been only 20 percent, equivalent to no more than 2 percent of total cost and 4 percent of value added.[*] The coke savings indicated by company X's data amounted to 5 to 10 percent of total coke consumption for each doubling of furnace size, depending on the initial furnace size and the quality of the iron-bearing materials. If actually realized, these savings would have reduced total cost by more than 1 percent, but less than 3 percent. Thus, a decision to

---

[*]Depending on plant location, it is also possible that the construction of larger furnaces saved land costs.

203

build one large rather than two small furnaces would have saved at least 2 percent but no more than 7 percent of total cost (or from 5 to 15 percent of value added). Rebuilding would have had similar effects on labor and coke costs (proportionate to the gain in size), but the savings in unit capacity costs are indeterminant.

The limited magnitude of these potential cost savings due to enlarged scale indicate that increases in the average scale of blast furnaces were largely dependent on growth in the demand for pig iron and shifts in regional demand. When the demand for pig iron in a given region grew slowly, increases in average furnace size occurred primarily by rebuilding and scrapping. The maximum savings in annual operating costs of 5 percent that could be achieved by replacing existing furnaces with new units twice as large would not have offset the annual depreciation and interest charges of the new facilities (based on unit capacity costs reported in Table 9.2), except perhaps during 1900-20.* But the annual rate of growth in demand during 1900-16 was high (6 percent); and there was a major shift in the location of industry capacity from the east and south to the Pittsburgh-Youngstown corridor and cities on the Great Lakes, so the growth rate in those locations was even higher. Hence, there was a high frequency of new furnace construction and a low frequency of scrapping during this period.

This earlier period stands in sharp contrast to 1916-29 (no growth in demand), 1930-39 (depressed demand), and 1941-70 (an annual growth rate of less than 2 percent). The frequency of new furnace construction was much lower and scrapping much higher after 1920, while rebuilding became an important source of additional capacity and increased scale. In fact, the slow growth in demand from 1955 to 1970 (25 percent) was exceeded by the increase in the capacity of the blast furnace stock resulting from improved materials, controls, and operating practice (50 percent).

The cost structure of the industry indicates that the major savings in cost have been achieved through technical improvements relating to materials and energy—improvements that are not dependent on the scale of furnaces except insofar as increased scale lowers coke rates. These improvements occurred in three areas, (1) reductions in the relative amounts of coke and iron-bearing materials blown out of the furnace as flue dust and increases in the proportion of flue dust

---

*This argument assumes that technical improvements made in the industry could be incorporated on existing furnaces at a reasonable cost (for example, while these furnaces were being relined or rebuilt), and that maintenance and repair costs did not rise rapidly as furnaces aged. To the extent that these assumptions are wrong, the savings in operating cost could have been far greater than 5 percent for replacing an old furnace with a new unit twice as large.

recovered, (2) reductions in the coke rate attributable to increased blast temperatures (accompanied by fuel and steam injection in recent years) made possible by improved firebrick quality and gas cleaning methods, and (3) the development of high quality iron-bearing materials—self-fluxing sinter and pellets—that lowered the coke rate enough to more than offset the higher costs of agglomeration (see Table 9. 7). To these three sources one could add that the transportation costs of raw materials as a percentage of product value were reduced greatly in the Pittsburgh-Youngstown corridor and in blast furnace plants located on the Great Lakes, from 40 percent in the 1930s to 30 percent in the early 1950s, and 20 percent in the 1960s (see Table 7. 5). This result, however, does not appear to have been the result of innovation in the transportation sector, but rather the result of a much slower growth in the costs of this sector compared with the blast furnace sector. Undoubtedly, the great age of the ore carriers operating on the Great Lakes and the railroad track beds contributed to the slow growth in transportation costs as these facilities became fully depreciated. On the other hand, the development of large scale ocean transport played an active role in reducing the cost of transporting foreign ore and agglomerates to plants located on the east coast and in the south, by approximately 40 percent (in current dollars) from 1954 to 1970.

The Present and Immediate Future

Between 1908 (the midpoint of the major expansion in new construction that occurred during 1900-16) and 1970, the average scale of actively utilized furnaces rose almost fourfold (measured by furnace hearth area) while total output rose threefold. In addition, average output per square foot of hearth area rose by 50 percent due to materials-related and technical improvements. Hence, output per furnace day doubled from 1908 to 1970 relative to the growth in total output; and, not surprisingly, the number of furnaces fell by almost 50 percent.

The doubling of the capacity of actively utilized furnaces relative to the expansion in total output greatly understates the extent to which developments in scale and technology have combined to increase the capacity of the largest and most technically advanced furnaces. In 1908, the hearth area of the largest furnace was only 50 percent greater and its capacity 70 percent greater than the corresponding figures for the typical utilized furnace. In comparison, in 1970 the hearth area of the largest furnace (located in Japan, with a hearth diameter of 45 feet) was almost three times as large and its capacity six times as great as the average utilized furnace in the United States. Thus, the capacity of the largest and most technically advanced furnaces has risen three and one-half times more rapidly than the capacity of an average-sized furnace and seven times more rapidly than the growth in total output since 1908.

The most modern furnaces are capable of producing almost four million tons of iron, and it appears likely that the minimum efficient plant size in modern, vertically integrated steel plants is now dictated by the scale of blast furnaces rather than by the rolling mills, as in the past.*

The cost advantage (if any) of these giant furnaces relative to the current average size furnace in the United States is not known. In 1972, the largest U.S. furnace had a hearth diameter of 38 feet— considerably larger than the average size furnace but considerably below the maximum. It is doubtful that the cost saving in the United States can be measured accurately until one of the domestic steel firms decides to construct one. It seems reasonable to claim, however, that the maximum possible savings in total unit costs (or value added) can be estimated by assuming that (1) the "two-thirds rule" applies to the relationship between investment and capacity (given the technical characteristics of the furnace) and (2) the labor force requirement per new furnace is constant. If this guess is correct, the maximum savings in total unit costs resulting from a decision to construct a huge new furnace (with a hearth diameter of 45 feet) rather than two furnaces with hearth diameters of 32 feet would be about 5 percent, equivalent to a 10 percent reduction in value added (including energy).† The cost saving could be much larger if energy consumption per ton of iron produced is lower on the larger furnace or if the larger furnace were to replace smaller furnaces in worn out condition.

It is unlikely that the slow growth in demand for pig iron will result in many occasions in the near future where a plant is faced with the decision on the best way to acquire four million tons of (additional) pig iron capacity. Hence, the most likely circumstances favoring the construction of large new units would appear to arise in plants with large numbers of furnaces (some of them small) in operation, where the capacity of a giant new furnace would replace, at least partially, the capacity of existing units.

Balanced against these suppositions are the more concrete facts pertaining to the advantage of pellets and high grade sinter over most other iron-bearing materials. Because pellets save coke and because

---

*This is particularly true when it is recognized that two blast furnaces per plant result in somewhat lower average costs than one furnace, ceteris paribus.

†These estimated savings assume that coke and injected fuels account for 30 percent of total cost, labor 4 percent of total cost, and capital charges (including "normal profit") 15 percent of the total cost of operating a blast furnace with a hearth diameter of 32 feet.

the price of energy materials has risen rapidly in recent years and probably will continue to rise, it seems more than likely that the domestic industry will continue to expand pellet production and consumption over the next few years. If this prediction is correct and if the demand for pig iron continues its slow advance, then the result will be to hasten the obsolescence of many small furnaces (depending on their location) as the number of utilized furnaces drops. This development would tend to reduce the likelihood that many very large new units will be constructed in the next few years because they would displace increasingly more efficient furnaces.

LITERATURE ON ECONOMIC THEORY PERTAINING TO SCALE

Bain, J. "Economies of Scale, Concentration, and the Condition of Entry in Twenty Manufacturing Industries." American Economic Review 44 (1954): 15-39.

Baumol, William. Economic Theory and Operations Analysis. Englewood Cliffs, N.J.: Prentice-Hall, 1961.

Chamberlin, Edward. "Proportionality, Divisibility, and Economies of Scale." Quarterly Journal of Economics 62 (1948): 229-62.

Chenery, Hollis B. "Engineering Production Functions." Quarterly Journal of Economics 63 (1949): 507-31.

Dean, Joel. Managerial Economics. Englewood Cliffs, N.J.: Prentice-Hall, 1951.

Dryden, C., and R. Furlow. Chemical Engineering Costs. Columbus: Engineering Experiment Station, 1966.

Enos, John. "Innovation in the Petroleum Refining Industry," in The Rate and Direction of Inventive Activity. Princeton: National Bureau of Economic Research, 1958.

Gold, Bela. "Economic Effects of Technological Innovations." Management Science 11 (1964): 105-34.

_____. Explorations in Managerial Economics. New York: Basic Books, 1971.

_____. Foundations of Productivity Analysis. Pittsburgh: University of Pittsburgh Press, 1955.

_____. "The Framework of Decision for Major Technological Innovation: Values and Research," in Values and the Future, edited by K. Baier and N. Rescher. New York: The Free Press, 1969.

_____. "Technology, Productivity and Economic Analysis." Omega 1 (1973): 5-24, 181-91.

Johnston, John. Statistical Cost Analysis. New York: McGraw-Hill, 1960.

Kaldor, Nicholas. "The Equilibrium of the Firm." Economic Journal 44 (1934): 60-76.

Knight, Frank. Risk Uncertainty and Profit. New York: Houghton Mifflin, 1921.

Lerner, Abba. Economics of Control. New York: Macmillan, 1944.

Mansfield, Edwin. "Speed of Response of Firms to New Techniques." Quarterly Journal of Economics 77 (1963): 290-309.

_____. "Technical Change and the Rate of Imitation." Econometrica 29 (1961): 741-66.

Moore, Frederick T. "Economies of Scale: Some Statistical Evidence." Quarterly Journal of Economics 73 (1959): 232-45.

Moroney, John R. "Cobb-Douglas Production Functions and Returns to Scale in U.S. Manufacturing Industry." Western Economic Journal 6 (1967): 39-51.

Robinson, Joan. The Economics of Imperfect Competition. London: Macmillan, 1933.

Salter, W.E.G. Productivity and Technical Change. Cambridge, England: Cambridge University Press, 1960.

Saving, Thomas R. "Estimation of Optimum Size of Plant by the Survivor Technique." Quarterly Journal of Economics, 75 (1961): 569-607.

Shuman, S., and S. Alpert. "Economies of Scale: Some Statistical Evidence: Comments." Quarterly Journal of Economics 74 (1960): 493-97.

Smith, Adam. Wealth of Nations, Book I. New York: Augustus M. Kelley, 1967.

Smith, Caleb. "Survey of the Empirical Evidence of Economies of Scale," in Business Concentration and Price Policy, Conference sponsored by the National Bureau of Economic Research. Princeton: Princeton University Press, 1955.

Solow, R.M. Minhas, K. Arrow, and H.B. Chenery, "Capital-Labor Substitution and Economic Efficiency." Review of Economics and Statistics 43 (1961):

Stigler, George. "The Economies of Scale." The Journal of Law and Economics 1 (1958): 54-71.

Walters, Alan. "Economies of Scale: Some Statistical Evidence: Comment." Quarterly Journal of Economics 74 (1960): 154-57.

_____. "Production and Cost Functions: An Econometric Survey." Econometrica 21 (1963): 1-61.

LITERATURE PERTAINING TO BLAST FURNACES

Anglo-American Council on Productivity. Productivity Team Report, Iron and Steel. London: Anglo-American Council on Productivity, 1962.

Babcock, D. E. "The Analysis of B. F. Operating Conditions by Means of the Theoretical Interpretation of the Top Gas and Iron and Slag Analysis." AISI Regional Technical Meeting, 1962.

Bogdandy, L. von, G. Lange, and P. Heinrich. "Entwicklungs Aussichten und Grenzen des Hochofens [Prospects of Development and Limits of the Blast Furnace]." Stahl und Eisen 88 (1968): 1177-88.

Chapman, H. H. Iron and Steel Industries of the South. Tuscaloosa: University of Alabama Press, 1953).

Collison, William H. "Natural Gas Injection at Great Lakes Steel Blast Furnaces." Iron and Steel Engineer, 39 (1962): 73-81.

Dailey, W. H. "Blast Furnace Performance—Pellets vs Sinter." Iron and Steel Engineer 40 (1963): 107-20.

Daugherty, C. R., M. G. DeChazeau, and S. S. Stratton. Economics of the Iron and Steel Industry. New York: McGraw-Hill, 1937.

Decker, A. "Les Injections de Combustibles Auxiliares dans le Harrts Fourneaux." Journees Internationales de Siderurgie, Luxembourg (1962): 213-22.

Doi, Y., and K. Kasai. "Blast Furnace Practice with Self-Fluxing Sinter Burden." Journal of Metals 11 (1959): 755-59.

Flint, R. V., "Effect of Burden Materials and Practices on Blast Furnace Coke Rate." Blast Furnace and Steel Plant 50 (1962): 47-58.

Graff, H. M., and S. C. Bouwer. "Economics of Raw Materials Preparation for the Blast Furnace." Journal of Metals 17 (1965): 389-94.

Himber, F., and Dutilloy. "Theoretical Aspects of High Top Pressure on Operation of the Blast Furnace." Journees Internationales de Siderurgie, Luxembourg (1962): 202-212.

Hogan, William T., S. J. Productivity in the Blast Furnace and Open Hearth Segments of the Steel Industry: 1920-1945. New York: Fordham University Press, 1950.

Johnson, Joseph E. Blast Furnace Construction in America. New York: McGraw-Hill, 1917.

Joseph, T. L. "The Potential and Limitations of High Blast Temperatures." Blast Furnace and Steel Plant 49 (1961): 239-46, 324-28.

MacDonald, N. D. "The Effect of Screened Sinter on Furnace Capacity." Proceedings: Blast Furnace, Coke Oven and Raw Materials Committee 20 (1961): 2-15.

Manners, Gerald. The Changing World Market for Iron Ore 1950-1980: An Economic Geography. Baltimore: The Johns Hopkins Press for Resources for the Future, 1971.

Marshall, W. E. "Taconite Pellets with Blast Furnace." Journal of Metals 13 (1961): 308-13.

Meinhausen, G. "Iron and Steel Works—Maximum Capacity, State of Planning, and Chances of Development." Stahl und Eisen 90 (1970): 153-61.

Melcher, N. B. "Bureau of Mines Use of Natural Gas in an Experimental Blast Furnace." AIME Blast Furnace, Coke Oven and Raw Materials Proceedings 18 (1959): 69-74.

Michard, J. "Etude Theorique de l'Injection de Fuel a Temperature de Vent Constante." Journees Internationales de Siderurgie, Luxembourg (1962):

Rice, Owen. "Three Blast Furnace Questions." Blast Furnace and Steel Plant 33 (1945): 1523-28.

Richards, E. "Pressure Blast Furnaces Show Greater Production, No Special Problems." Blast Furnace and Steel Plant 36 (1948):

Strassburger, J. H. , D. C. Brown, R. L. Stephanson, and T. E. Dancy. Blast Furnace--Theory and Practice. New York: Gordon and Breach Science Publishers, 1969.

United States Steel Corporation. The Making, Shaping and Treating of Steel. (Pittsburgh: U.S. Steel Corporation, various issues).

MAJOR SOURCES OF DATA

American Iron and Steel Institute, Directory of Iron and Steelworks In America (New York: AISI, all editions from 1900).

American Iron and Steel Institute, Annual Statistical Report (New York: AISI, all issues).

American Iron Ore Association, Iron Ore (Cleveland, AIOA, all issues since 1961).

U. S. Bureau of the Census, Census of Manufactures (Washington, D. C.: U.S. Department of Commerce, all issues since 1899).

U. S. Bureau of Mines, Minerals Yearbook (Washington, D. C., U. S. Dept. of the Interior, all issues since 1950).

malleable pig iron, 46 (see also, pig iron products)

Manners, Gerald, 171

Mansfield, Edwin, 22

market structure, 21-27: competitive, 4-5, 11, 18, 25, 26, (see also, equilibrium)

materials, measurement of, 23; and power rule, 11; theoretical treatment of quality changes in materials in production models, 19, 23, 31-33, 36-44

Meinhausen, G., 129, 131, 132, 171

Moore, F.T., 9-11

natural gas, 86, 108-111 (see also, energy, fuel injectants)

natural ores, 45, 69, 185; sources, 52-53; types, 50, 55 (see also, concentrated ores, iron-bearing materials, ores)

ores, consumption during 1955-70: in blast furnaces, 80-83, 84, 97; in sinter plants, 73-77; effect of quality changes on: blast furnace productivity, 117-119, coke flux rate, 90-91, 92-95; quality of, compared to sinter, 54-56 data on, 79 (see also, iron-bearing materials)

output, products, 1-3, 5-6, 23-24; broad definition of, 39-40, 41-42, measurement of, 23; in petroleum refining, 21 (see also, product mix)

overhead and profit, in the blast furnace sector, 181, 182-184; per ton of pig iron products, 192, 196, 200 (see also, costs as value-added)

pellets, 52-53, 54-55, 67, 91, 97; consumption of: absolute quantities, 80-83, 84; relative quantities, 100; effect on: blast furnace productivity, 115, 117, 119-

121, 124, 126-127, 131-132, 185, 205; and flux rate, 100-103, 103-106, 107, 111-113, 204-205;· (see also, agglomerates, iron-bearing materials, sinter)

pig iron products, 45-46; correlation with furnace size, 106-107: iron content of, 56; rate of production, 60-69 (see also, basic iron, Bessemer iron, ferroalloys, foundry iron, malleable iron)

plants, 1-2, 3, 4-5, 7, 15-17, 18-21, 21-22, 24-25, 26-32; applicability of power rule to, 9-11; extent of horizontal/vertical integration and measurement of scale, 31, 33, 34-35, 36 (see also, capital, production functions)

power rule, also six-tenths rule, 9-11, 171, 203-204, 206

prespecified production functions (see, production functions)

prices: of blast furnace inputs, 172, 174, 176, 186: composite index, 180-181, 191, 196-197: of coke, also, coking coal, 172-174, 177, 181-183, 185, 186-189, 200: of flux, 176-177; of inputs, 2-4; and notes, 12, 24, 40; effect of changes in, 15-16, 17, 18-19; of iron-bearing materials, composite index, 182, 185-188, 191-200; of labor services (see, wages): manufactured commodities, also wholesale price index, 180, 185-187, 196-200; of ores, 174-177; of output, 4, 5, 18, 24, 25; of pellets, 174-177· of pig iron products, composite index, 180-181, 182, 191-192, 196-198

product mix, and measures of plant size, 15, 16-17, 18-19, 22: and production models, 31, 35, 36, 37-43: and quality changes due to innovations, 23, 25-26 (see also, output)

production functions, production

217

unit costs (see, costs)
utilization rate: of capacity (see, capacity, utilization)

value-add (see, costs as value-added)
variable costs (see, costs, variable)
vertical integration of activities in plants (see, capital, materials, plants, production functions)

wages, in the blast furnace sector and basic steel, 145-147, 172-174, 177, 182-185, 189-190, 200
Walters, Alan, 11
Womer, R., 9

MYLES G. BOYLAN, Jr. is an assistant professor in the department of economics, Case Western Reserve University, and research associate in the Research Program in Industrial Economics of Case Western Reserve University in Cleveland, Ohio, where he received his Ph.D.

PLANT SIZE, TECHNOLOGICAL CHANGE, AND
INVESTMENT REQUIREMENTS: A Dynamic Frame-
work for the Long-Run Average Cost Curve
David Huettner

INDUSTRIAL LOCATION DECISIONS IN DETROIT:
A Comparison with Chicago and Atlanta
Lewis Mandell

THE JAPANESE STEEL INDUSTRY: With an Analysis
of the U.S. Steel Import Problem
Kiyoshi Kawahito

HIGH LEVEL MANPOWER
Dale H. Hiestand